CALIFORNIA LANDSCAPES

TOURING NORTH AMERICA

SERIES EDITOR
Anthony R. de Souza, *National Geographic Society*

MANAGING EDITOR
Winfield Swanson, *National Geographic Society*

CALIFORNIA LANDSCAPES

*Los Angeles, Big Sur,
San Francisco, Yosemite, and
Death Valley*

BY
**MARTIN S. KENZER
DOUGLAS J. SHERMAN
ROBERT A. RUNDSTROM**

AND
BERNARD O. BAUER

with contributions by
DAVID HORNBECK, PAUL F. STARRS,
WILLIAM K. CROWLEY, and JOHN WOLCOTT

RUTGERS UNIVERSITY PRESS • NEW BRUNSWICK, NEW JERSEY

This book is published in cooperation with the 27th International Geographical Congress, which is the sole sponsor of *Touring North America*. The book has been brought to publication with the generous assistance of a grant from the National Science Foundation/Education and Human Resources, Washington, D.C.

Rutgers University Press
109 Church Street
New Brunswick, New Jersey 08901

The paper used in this book meets the minimum requirements of American National Standard for Information Sciences—Permanence of Paper for Printed Library Materials, ANSI Z39.48-1984.

Library of Congress Cataloging-in-Publication Data

California landscapes: Los Angeles, Big Sur, San Francisco, Yosemite, and Death
    Valley / by Martin S. Kenzer ... [et al.]; with contributions by David Hornbeck, Paul
    F. Starrs, William K. Crowley, and John Wolcott.
        p.   cm.—(Touring North America)
    Includes bibliographical references and index.
    ISBN 0-8135-1886-5 (cloth)—ISBN 0-8135-1887-3 (paper)
    1. California—Tours.   I. Kenzer, Martin S., 1950–   . II. Series.
F859.3.C27   1992
917.9404′53—dc20   92-10531
                                                                          CIP

First Edition

Frontispiece: The famous 1953 four-level Stack interchange at the intersection of the Hollywood (upper), Harbor (left), and Pasadena (right) freeways in Los Angeles. Photograph with the kind permission of the CALTRANS Library History Center.

Series design by John Romer

Typeset by Peter Strupp/Princeton Editorial Associates

△ Contents

△ Foreword

Touring North America is a series of field guides by leading professional authorities under the auspices of the 1992 International Geographical Congress. These meetings of the International Geographical Union (IGU) have convened every four years for over a century. Field guides of the IGU have become established as significant scholarly contributions to the literature of field analysis. Their significance is that they relate field facts to conceptual frameworks.

Unlike the last Congress in the United States in 1952, which had only four field seminars all in the United States, the 1992 IGC entails 13 field guides ranging from the low latitudes of the Caribbean to the polar regions of Canada, and from the prehistoric relics of pre-Columbian Mexico to the contemporary megalopolitan eastern United States. This series also continues the tradition of a transcontinental traverse from the nation's capital to the California coast.

California embraces the greatest diversity of physical and cultural landscapes in America. It ranges from Death Valley, which is below sea level, to Mount Whitney, the highest peak in the lower 48 states, and from the most arid to one of the most rainy spots on the continent. The state extends more than halfway from Mexico to the Canadian border and includes some of the nation's most barren lands as well as some of the most complex and productive agricultural and industrial landscapes. No other state exceeds the richness of California's history and scenery, and the variety of its ethnic components. The character of California is revealed by four talented scholars: Professors Martin S. Kenzer, Douglas J. Sherman, Robert A. Rundstrom, and Bernard O. Bauer. In addition, the book gains from the contributions of four other geographers: Professors David Hornbeck (for numerous historical and contemporary vignettes), Paul F. Starrs (for the Gold Country tour from Yosemite to Mono Lake), William K. Crowley (for the Wine Country tour), and John Wolcott (for the walking tour of downtown L.A.).

Anthony R. de Souza
BETHESDA, MARYLAND

△ Acknowledgments

We acknowledge the dedicated work of the following cartographic interns at the National Geographic Society, who were responsible for producing the maps that appear in this book: Nikolas H. Huffman, cartographic designer for the 27th IGC; Patrick Gaul, GIS specialist at COMSIS in Sacramento, California; Scott Oglesby, who was responsible for the relief artwork; Lynda Barker, Michael B. Shirreffs, and Alisa Pengue Solomon. Assistance was provided by the staff at the National Geographic Society, especially the Map Library and Book Collection, the Cartographic Division, Computer Applications, and Typographic Services. Special thanks go to Susie Friedman in Computer Applications for procuring the hardware needed to complete this project on schedule.

We thank David J. Larson, California State University, Hayward, for help field-testing part of the tour. For assistance with illustrations and permission to use their photographs, we are grateful to: Caltrans Library History Center, Dennis McClendon, *Planning* magazine, Los Angeles Department of City Planning, Los Angeles Convention and Visitors Bureau, Elliot McIntire, Barbara Kopel, Stephen Maikowski, Adrian Atwater, Cam Sutherland at the University of Nevada Press, United States Geological Survey, Al Glass.

We thank Lynda Sterling, public relations manager and executive assistant to Anthony R. de Souza, the series editor; Richard Walker, editorial assistant at the 27th International Geographical Congress; Natalie Jacobus and Cynthia Suchman, who proofread the volume; and Tod Sukontarak, who indexed the volume and served as photo researcher. They were major players behind the scenes. Many thanks, also, to all those at Rutgers University Press who had a hand in the making of this book—especially Kenneth Arnold, Karen Reeds, Marilyn Campbell, and Barbara Kopel.

Errors of fact, omission, or interpretation are entirely our responsibility; the opinions and interpretations are not necessarily those of the 27th International Geographical Congress, which is the sponsor of the field guide and the *Touring North America* series.

PART ONE

Introduction to the Region

⚠ Introduction

Approximately 30 million people inhabit California, more people than in the entire country of Canada, and more than one tenth of the entire population of the United States. Similarly, if it were an independent country, because of its extraordinarily diverse economy, and the huge dollar value of the goods and services sold in California, its gross national product (GNP) would place it among the top ten countries. In sum, the "golden state" has one of the most varied and remarkable cultural landscapes in North America.

The place to monitor national trends is California. (The saying should be: As California [not Iowa] goes, so goes the nation. Iowa holds the nation's first caucus before the presidential election; the winner will, they say, win the primary.) This is true, among other reasons, because this state is far more than a great hub of activity: its population is massive—30 million and growing; it has been and remains the destination of immigrants (from all other states as well as from countries around the world)—at first from the east coast, but currently from every state as well as from all parts of Latin America, Asia, and elsewhere; its physical diversity and strong economic underpinnings endow it with qualities more like a country than a single state; and because of the above reasons and others still, it receives disproportionate attention from the news media and is therefore in our collective minds more than most other states or places.

Over time, the state has taken on an almost mythical image. This has evolved to the point where California—and many things presumed (rightly or wrongly) to be Californian—symbolizes specific things to many people. There is, therefore, a stereotypical view of the state, half true and half false.

California Landscapes

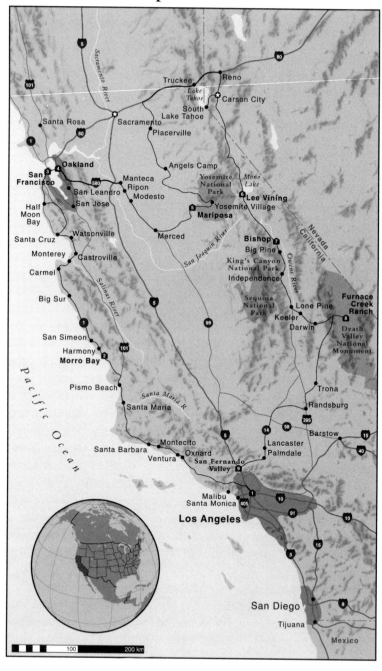

This field guide looks at a representative cross-section of the state. California is the third largest of the fifty states (behind Alaska and Texas), more than 156,000 square miles. Be warned! Even native Californians can underestimate the size of this state and the time it takes to get from place to place. We have divided the itinerary into nine segments, each of which can, in theory, be accomplished in a single day of long, hard driving (and no traffic jams). But you will enjoy your tour of California much more if you plan to take at least two days for each segment of the journey and stop often to explore places by foot, sample the food, and meet the people.

THE PEOPLING OF
 CALIFORNIA

THE INTERIOR

The arrival of the Europeans profoundly altered the human geography of the state, especially in reference to settlement patterns and the ever evolving configuration of land occupance. Californians have always seen the state divided into a northern half and a southern half. Even during the initial Spanish settlement, there was a distinct geographical split between the northern and southern portions of the state. This trend persists to this day—perhaps even more so than in the past—to the point where a majority of all tourist-oriented guidebooks appropriately split California into two volumes. During the past fifteen years, however, the state has begun to develop yet a third geographic region—an agricultural interior—that runs from Bakersfield north to Sacramento and beyond.

This emerging "interior California" extends from the Tehachapi Mountains over the San Joaquin Valley, northward through the Sacramento Valley, until it encounters the subalpine portion of the state that forms California's northern edge. For the most part, this region has taken a predictably secondary role in shaping California's image. Spanish and Mexican settlement during the eighteenth and nineteenth centuries was coastal, and the standard of a predominately coastal pattern continued well into the American period. During the Gold Rush of the mid-1800s, Marysville and Sacramento functioned as gateways to the gold fields, but bona fide urban settlement grew up along the coast.

Deliberate, systematic settlement of the interior began in the 1870s, spurred by the advancing railroad. Today, however, the interior supports deep-water ports at Stockton and Sacramento, and considerable urban growth is occurring in El Dorado, Placer, Sacramento, and Yolo counties. This area, centered on Sacramento, is now considered by many to be one of the most dynamic metropolitan areas in California, with a population of nearly 2 million. Fresno and Bakersfield, in the central and southern portions of the Great Central Valley, have likewise experienced considerable urban growth, and both have become large cities in their own right.

Notwithstanding the extraordinary city growth, this segment of the state is, and will continue to be, dominated by agriculture. The Central Valley's 15 million acres—an area equal in size to the entire territory of New Jersey, Massachusetts, and Vermont, combined—holds three fourths of California's irrigated cropland, and a full one sixth of the total irrigated cropland of the United States.

The emergence of California's burgeoning interior dramatically illustrates the present-day unfolding of the California dream: The very same motivations that coaxed settlers to the coastal region of the state—the chance to own one's home, a better paying job, an enhanced life-style, especially for one's children—now lure people into California's less enticing interior.

URBAN GROWTH

Unlike other frontiers in the United States, California had developed a large urban population by the 1850s, in part because towns played an integral role in the state's early Gold Rush settlement. The Gold Rush flooded California with towns of all sizes and launched a period of rapid urban growth. In 1860, 20 percent of California's population resided in an urban place. In comparison, during a comparable period of intense settlement, only 1 percent of Ohio's population lived in urban places. By 1870—only twenty years after achieving statehood—California was among the most urbanized states in the country. To say that Californians have shown a strong preference for living in cities is an understatement. Since 1940, California's population has more than tripled, and most of the increase has occurred in coastal areas, especially in and around San Francisco and Los Angeles. In 1900, the San Francisco and Los Angeles metropolitan areas held 49 percent of the state's population. That number increased to 65 percent in 1930, but by 1990 it had declined slightly to 62 percent. The change in population density since 1940 reflects the ever-increasing concentration of the urban population along the coast and toward the southern portion of the state. Today, California's population is highly concentrated: More than three fourths of all residents inhabit a mere 1 percent of the total land area. Moreover, the state is grossly unbalanced between northern and southern California. Since the 1950s, southern California has received the bulk of the new growth; approximately six and a half of every ten new arrivals made southern California their home. During the 1980s, for example, southern California's population increased by over 20 percent, and the area south of Santa Barbara now contains 19 million of the state's roughly 30 million people—the vast majority of them living in an urban locale.

Along with an increase in total urban population, the number of urban locations in the state proliferated also. At the turn of the century, California had only forty urban places (population, 2500 or more). By 1940, that number had increased fourfold, and since 1940 the number of urban places has more than tripled, to reach over 600 today.

The most important change in the distribution pattern of California's population in recent years has been occurring in metropolitan areas. There has been a rapid movement from the central cities to the built-up areas surrounding them. In general, the larger the city, the more pronounced this movement has been. The net effect of this pattern has been to reduce the proportion of the population that lives in the central city and thereby to increase the proportion living in the smaller cities.

CALIFORNIA'S NATIVE POPULATION

American Indians comprise contemporary California's smallest population, amounting to just 0.8 percent of the state's total population. This can be misleading, however, because only Oklahoma had as many Indians—250,000—in the 1990 Census. This figure, and the diverse and dispersed nature of this ethnic group, bears striking similarities to the condition of California Indians before European contact.

The diversity of California's physical environments is matched by the complex mosaic of indigenous cultures that thrived here before European contact in the eighteenth century. In fact, the variety of indigenous cultures exceeded that for any other part of North America. Six major language stocks (four along the North Coast alone), 80 languages, and more than 300 dialects were spoken here.

California was as appealing to indigenous people as it is to immigrants today. Thirteen percent of the pre-European indigenous population living in the area that is now the United States lived in California. Coincidentally, about the same percentage of U.S. citizens lived in California in 1990!

Dry summers, rugged terrain, and poor soils virtually eliminated the possibility of early non-irrigated agriculture, except along the Colorado River. As ever, California's physical environment provided ample local resources, however, and extensive statewide and regional trade networks provided early Indians with a quite varied diet (just as Californians enjoy today).

European contact unleashed effects similar to those experienced by all native people in North America, with few exceptions. The Spanish missionaries (1769 to 1836) saw Indians as heathen "children" to be converted, exploited, and absorbed into their empire. Eighty-five thousand Indians were christianized during the sixty-

CALIFORNIA INDIAN POPULATION

For almost 200 years descriptions of the California Indians were routinely negative. First Spanish, then Mexican, and later American settlers portrayed them as culturally deficient— a group perched on the lowest rung of the civilization ladder. The usual signs of an advanced culture were not evident to the settlers: agriculture was nowhere to be found; and only the weakest socio-political organization was noticeable, and then only in a few, small, scattered groups. Because what was observed was not thought to be significant, precious little of the Indians' languages, institutions, material culture, or even the size and location of the various groups, was recorded before they began to fade from the scene in the wake of European settlement.

The inaccurate descriptions and negative attitude toward the California Indians stem in part from the inability of early observers to differentiate among the hundreds of small individual Indian groups, and to understand and appreciate the manner in which they made use of the environment. The California Indians had developed sophisticated social and economic institutions rather than technological skills to cope with a remarkably diverse and complicated natural landscape. Within this environmental expanse of mountains and deserts, wet winters and dry summers, grasslands and oak forests, narrow valleys and flat plains, Indian groups took advantage of every available local resource. In turn, each stretch of beach, valley, and hillside had its own contingent of Indians, often very different from those living only a short distance away. The California Indians exhibited a bewildering mosaic of regional differences in human occupancy. But these differences were not apparent to the intruders because the standard cultural signposts were not evident. As a result, the subtle balance maintained between Indian and environment went unnoticed, completely lost in the clamor to save their souls and take their land. Instead, the Indians were grouped together, labeled cultural laggards, and dismissed as crude, simple, and primitive. The more earnest and serious population estimates range upward from 133,000. The tendency in the past thirty years has been to revise the estimates upward—as has been true for the entire native population throughout all of North America. The largest estimate suggests that approximately 310,000 existed in California at the time of European settlement.

In general, the population was concentrated along the coast, diminished toward the interior, and dropped rapidly after crossing the Coast Ranges. The native population was generally thinly distributed in the more arid regions and almost nonexistent in mountain areas above 5,000 feet. Overall, population density varied considerably, ranging from less than one person per square mile to more than eleven per square mile. Large villages tended to cluster in three areas: along the Santa Barbara Coast, in the foothills of the upper San Joaquin Valley, and along the lower reaches of the Sacramento Valley.

The patterns suggest that California had a very large population. Today—except for the most remote areas of northern California and the desert areas of southern California—the California Indians are almost nonexistent. Both Spanish and early Anglo settlement reduced the Indian population considerably. By 1850 (the ostensible start of Anglo settlement), the number of native Californians had declined to about 170,000. Thirty-five years later, the number of California Indians listed in a special census numbered only 15,000.

five year reign of the missions; the vast majority then typically died from epidemic disease.

The rest of the nineteenth century brought a litany of abuses well-known to North Americans: slave labor, extermination, treaty, and removal. Three events are worth noting as uniquely Californian in this history. The Russians were the first to enter into a treaty with an Indian group here (the Pomo in 1812). Later, the 1850s Gold Rush led to genocidal campaigns carried out by vigilante mining groups in the rugged eastern and northern portions of the state. Additionally, treaties with the U.S. government remained unratified and hidden in U.S. Senate archives so that gold prospectors could cross reservation boundaries with impunity. Finally, the so-called Modoc Wars of the 1870s are infamous in California because they were the last violent confrontation between Indians and white settlers, thereby "closing" the state's frontier. In the winter of 1872 to 1873, 150 Modoc battled 1,000 U.S. Army soldiers in the isolated lava plateaus of northeastern California. The Modoc were captured and shipped off to Oklahoma Indian Territory, but not before their leaders were executed and their heads sent to the Smithsonian Institution for "scientific" investigation. By 1900, the California Indian population numbered just 15,000.

There has been a resurgence in both the Indian population and Indian cultural awareness since 1900, primarily in the aftermath of World War II. The single greatest factor in this growth has been the influx of Indians from other states, notably from Oklahoma and reservations in the Dakotas and the Southwest. In the 1960s, federal policy mandated termination of all reservations and the relocation of Indians to urban centers. Los Angeles, San Francisco, and Oakland became destinations for Indians across the country. By the 1980s, more than half of all relocated Indians in the United States had resettled in these three cities. Important health and human resource centers—including an Oakland center for training tribal attorneys and resource managers—have existed in these cities for some time. Now, the vast majority of Indians in the state live here.

California has no fewer than 101 different reservations and "rancherias," far more than any other state. Numerous small reservations cluster in two main areas: San Diego County and central Riverside County, and along the north coast in Humboldt and Mendocino counties. The far northern part of the state contains most of the rancherias, another unique California phenomenon. A rancheria is a "reservation-in-miniature," often consisting of a cemetery or a single home on just a few square miles or a quarter of an acre. Although outwardly unremarkable, these sites are federally recognized Indian land that provide a crucial land "anchor" and meeting place for nearby Indian residents. Ongoing demonstrations and legal battles for resources, especially fishing rights, are frequently staged on north coast rancherias.

"Indianness" has always been esteemed by a large portion of the white population in California. The origin of this attitude lies primarily in the perception that Indians seek a harmonious relationship with nature, an equally desirable goal for many non-Indians in the state. A growing pride in Indian heritage, no matter how dilute, and the removal of federal legal restrictions on the definition of "Indian" have also contributed to the resurgence in Indian culture. To claim an Indian heritage or to study Indian ways has a certain cachet in California now, and it seems to be spreading eastward.

SPANISH SETTLEMENT OF CALIFORNIA

Having established two small military outposts (San Diego and Monterey), Spain began by 1770 developing other settlements in California. The plan was to found five missions along the California coast that would enforce Spain's claim to the entire Pacific Coast, neutralize the colonization efforts of other European countries, and, most importantly, protect its lucrative galleon trade with the Philippines. The initial thrust of Spanish colonization centered on two time-tested frontier institutions, the mission and the

Cattle ranch in San Luis Obispo County. Photograph by Paul F. Starrs.

presidio. A third settlement institution, the pueblo, was brought to California in 1777. All three frontier institutions were to work in tandem, each performing a specific role in Spain's colonization efforts.

Spain's ventures in California depended on a settlement strategy that included absorbing the indigenous population. To effect permanent settlement, Spain employed a system of Catholic mission stations that converted the local inhabitants—voluntarily—to Christianity and trained them to become loyal Spanish subjects.

Presidios formed the defensive arm of Spanish settlement in California. As an agent of the government, the presidio was responsible for defending the coast, subduing hostile Indians, maintaining peaceful relationships with friendly Indians, and acting as the secular authority until a civil government could be established. Four presidios were founded on the California coast—San Diego

Authentic Spanish colonial architecture: Mission Santa Barbara. Photograph by Dennis McClendon.

(1769), Monterey (1770), San Francisco (1776), and Santa Barbara (1782)—each with a specific district to control. Each presidio was a military fort staffed by infantry and cavalry who were reassigned, in part, to duty at the missions within a particular presidio's jurisdiction. The mission–presidio system had been successful in settling other areas of the Spanish empire, and its initial deployment on the California coast was to differ little from previous practice.

Pueblos (civil communities) were a later addition to Spain's efforts to colonize California. They were established to supply the military with agricultural products, thereby reducing the high transport costs incurred by shipping food from Mexico and ensuring a consistent food supply in California. Pueblo citizens were also to set an example of Spanish life for the Indian to follow, and to act as a reserve militia in times of emergency. Only three pueblos were founded in California—San Jose (1777), Los Angeles (1777), and Santa Cruz (1797)—and their success was less than expected. The pueblos never seemed to produce the needed agricultural surplus, and many of the colonists were malcontents or criminals banished to California from Mexico.

Missions were founded (between 1769 and 1823) primarily in areas that contained large numbers of Indians, and they were allowed to take up and use as much land as was necessary to properly care for Indian converts. The missions could thus take advantage of good sites and Indian labor to expand into large, well developed estates and eventually dominate the secular settlements. At their height (1822), more than 22,000 Indians resided on mission property. Today, the twenty-one California missions are primarily tourist attractions: Missions Santa Barbara (founded in 1786) and San Juan Capistrano (1775–76) each account for over 300,000 visitors a year. Of all of the Spanish settlement institutions brought to California, the missions became the most dominant, and their locations form the basis of California's present-day urban structure. Mission locations were used as building sites for American cities, which currently account for almost 60 percent of the state's population.

IMMIGRATION AND ETHNICITY

European settlers were first drawn to California in large numbers by the unparalleled chance of digging a fortune from the gold fields. Others followed to take advantage of economic opportunities in the rapidly expanding cities or for the chance to obtain land. Real and imagined opportunities have made California one of the most diverse areas in the world in terms of the composition of its population.

Foreign-born immigrants have always been an important component of California's population. In 1860, foreign-born residents accounted for 39 percent of the state's total population. By 1900, the proportion of foreign born had declined to 28 percent and by 1990, to approximately 14 percent.

Since 1850, great changes have occurred in the number and relative importance of various "foreign-born" groups. The largest foreign-born populations to arrive during the Gold Rush were the Mexicans, the English, the Germans, the French, and the Irish. Together, these five groups accounted for 80 percent of the state's foreign-born citizenry. During the 1850s, however, immigrants from Asia began to arrive in large numbers, most notably the Chinese. By 1860, the Chinese had become the largest foreign-born group in California, numbering more than 35,000. Through 1890, the Irish and Chinese were ranked as the first- and second-largest foreign-born group in California.

The Chinese Exclusion Act, which limited the number of Chinese immigrants, went into effect in 1882. By 1900, with both the Chinese and the Irish on the decrease, Germans became the most numerous group in 1900 and 1910. Mexican immigrants to California began to swell in numbers after 1900, and by 1920 they had become the largest foreign-born group, a position they continue to hold.

The composition of California's population is rather exceptional because in no other area in the United States are blacks, whites, Asians, Mexicans, and Native Americans found together in such significant numbers. Indeed, minority groups represent a significant portion of California's present-day population. In 1990, for instance, blacks, Asians, Native Americans, Hispanics and a host of smaller groups accounted for approximately 47 percent of California's 30,000,000 residents, with those of Hispanic origin comprising 21 percent of the state's population. These groups are not evenly spread throughout the state, however; rather, they are concentrated in the eight counties that make up southern California. Moreover, while southern California contained 58 percent of the state's population in 1990, it was home to 67 percent of minority groups statewide—specifically, 63 percent of the black population and 71 percent of Spanish origin.

PHYSICAL CALIFORNIA

GEOMORPHIC PROVINCES OF SOUTHERN AND CENTRAL CALIFORNIA

The state of California is divided into eleven geomorphic, or physiographic, provinces. The field trip described in this volume visits seven of them. Following our clockwise circuit, we will cross parts of the Peninsular Ranges province, then the Transverse Ranges, Southern Coast Ranges, Great Valley, Sierra Nevada, Basin Ranges, the Mojave Desert, and then back across the Transverse Ranges to the Peninsular Ranges again.

The Peninsular Ranges

We will see only the northern limit of the Peninsular Ranges province in and around the Los Angeles Basin. The Peninsular Ranges, including the San Jacinto Mountains at the eastern edge of the L.A. Basin, are the northern extensions of the granitic spine of Baja California. San Jacinto (10,805 feet) is the highest peak in the range, and in winter it frequently provides a snowy backdrop for sun-seekers in Los Angeles and Palm Springs. The major geologic feature of the Peninsular Ranges is the Southern California batholith, comprising a large number of plutons associated with most of the major peaks in the ranges.

Physiographic Regions of California

Northern Coast Ranges

Sacramento River

Sierra Nevada

Lake Tahoe

Basin Ranges

Sacramento Valley

San Pablo Bay

San Francisco Bay

Great Valley

Nevada California

Mono Lake

Diablo Range

San Joaquin R

White Mtns

Monterey Bay

Southern Coast Ranges

Salinas Valley

Santa Lucia Range

Salinas River

San Joaquin Valley

Mt Whitney 4784m

Inyo Mtns

Death Valley

Panamint Range

Santa Maria R

Mojave Desert

Santa Ynez Mtns

Transverse Ranges

San Gabriel Mtns

San Bernardino Mtns

Santa Monica Mtns

Los Angeles Basin

Channel Islands

Peninsular Ranges

San Jacinto Mtns

Pacific Ocean

100   200 km

The major geomorphic feature of the Peninsular Ranges, in the vicinity of this field trip, is the Los Angeles Basin (Day Two itinerary) at the junction with the Transverse Ranges. Other characteristic landform assemblages are south of our area of interest.

The Transverse Ranges

The Transverse Ranges were so named because they trend east–west, across the grain of the California landscape. The field trip crosses the Transverse Ranges twice, in the western, then the central reaches (Days Two and Nine itineraries). The principal landforms of the western Transverse Ranges include the northern margin of the Los Angeles Basin, the Santa Monica and Santa Ynez mountains, and the San Fernando Valley, Oxnard Plain, and Santa Ynez Valley. Major rivers (all relatively small) include the San Gabriel, Los Angeles, Santa Clara, and Santa Ynez. All of this portion of the province is tectonically active, with extensive faulting and folding evident.

The deeply dissected mountains inland of Ventura and Santa Barbara remain relatively pristine because of their inaccessibility. For this reason, California's Condor Reserve is located therein. These ranges are also extremely prone to hazards associated with flooding, wildfire, and slope failures.

The central portions of the Transverse Range are encountered on the return to Los Angeles from Death Valley, and after leaving the Mojave Desert (Day Nine itinerary). South of Palmdale, we cross the San Andreas Fault, with Sierra Pelona to the northwest and the San Gabriel Mountains to the south. Farther east lie the San Bernardino Mountains, with a maximum elevation of 11, 502 feet at San Gorgonio Peak. The San Gabriel Mountains are an extremely active system, with rapid uplift and denudation rates. One result is some of the steepest average slopes in California.

The Sierra Pelona is mainly schists, whereas most of the exposed rocks in the San Gabriel Mountains are granitic. However, at several locations in the San Gabriels coarse marine sedimentary rock outcrops occur, including the hogback ridges of Vasquez

Clouds form over desert as convection currents lift hot air up from the dark soil. Photograph by Bernard O. Bauer.

Rocks, near Acton. The field trip route then follows the Soledad Basin, filled with Oligocene non-marine sediments, and including borax deposits, through the San Gabriel Mountains and back into the San Fernando Valley.

The Coast Ranges

The Coast Ranges parallel the California coastline northward from Santa Maria to the Klamath River. This geomorphic province is frequently subdivided into Northern and Southern Ranges that break at San Francisco Bay. We focus on the Southern Ranges (Days Two, Three, Four itineraries).

The major geomorphic features of the Southern Coast Ranges are, from west to east, the Santa Lucia Mountains, the Salinas

Rock stacks, Big Sur coastline. Photograph by Paul F. Starrs.

Valley, and the Diablo Range. The Gabilan Range lies east of Monterey Bay, and the Santa Cruz Mountains extend from Monterey Bay north to San Francisco. The largest rivers include, at the southern boundary of the province, the Santa Maria, then the Salinas River, and the Pajaro River.

Overall elevations of peaks in the Southern Coast Ranges are not especially high, the highest being Big Pine Peak at 6,828 feet. However, local terrain may be very steep, as evidenced by the cliffs rising from the Pacific along the Big Sur coast. Elevations exceeding 5,000 feet can be encountered within 4 miles of the ocean, at Cone Peak (5,155 feet). The present structure of this range results from extensive faulting and folding along the margins of the plate boundaries. Most of the Southern Coast Ranges lie west of the San Andreas rift zone.

Other major features of this province include the magnificent bays around Oakland: San Francisco, San Pablo, and Suison. The system of bays absorbs the drainage from the interior of the state, mainly via the Sacramento and San Joaquin rivers. The bays themselves result from a structural depression that has been greatly filled through both marine and fluvial sedimentation.

The Great Valley

The Great Valley comprises the Sacramento Valley (in the north) and the San Joaquin Valley (in the south) (Day Six itinerary). The Valley, approximately 450 miles long, is a partially filled structural trough. Down-warping of the valley was associated with Coast Range orogenies (mountain building) and westward tilting of the Sierra Nevada–San Joaquin blocks, beginning about 60 million years ago. Subsequently the province was a large inland sea that gradually filled via marine and fluvial sedimentation. The region remains prone to flooding because of its broad surface, low slope angles, and proximity to Sierran rainfall and snowmelt. In the 1860s an area approximately the size of Lake Michigan was inundated by intense runoff from the mountains surrounding the Great Valley.

The benefit of long-term sedimentation has been the development of extremely fertile soils, supporting a rich, diverse agricultural economy. At the same time, the long history of accretional infilling has resulted in the development of a large, relatively undistinguished surface. There are few major geomorphic features in the Great Valley. The largest rivers in the southern and central Great Valley drain into San Francisco Bay. They include the Kern, Kings, San Joaquin, Merced, Tuolumne, Stanislaus, and Sacramento rivers.

The Sierra Nevada

The Sierra Nevada Province is about 400 miles long. and as wide as 80 miles. The province includes the granitic bastion, Mount Whitney (14,496 feet), its highest point, and other fantastic land-

Yosemite Falls, Yosemite National Park. Photograph by Paul F. Starrs.

scapes responsible for the region's renown. In simple terms, the province is a mountain range composed of granitic intrusions, asymmetrically tilted westward. Consequently, the western slopes are relatively gentle, whereas the eastern slopes rise abruptly from the eastern valleys, with relative elevations exceeding 10,000 feet above the Owens Valley. Here, we emphasize the central portions of the range (Days Seven and Eight itineraries).

Most of the crustal movement associated with raising the Sierra Nevada occurred along the Sierra Nevada fault system that runs along the eastern edge of the range. Vertical movement within this system has exceeded an estimated 10,000 feet. Most of this movement has occurred in the last 3 million years, and the system remains active. The resulting high altitudes along the crest of the range served as foci for glacier formation during the Pleistocene. The erosion caused by movement of these glaciers carved many deep, scenically dramatic valleys on both sides of the range, including, of course, Yosemite Valley.

Remnants of the once powerful glaciers can be found in the Lyell and Palisade glaciers. Many of the Sierran valleys were once filled with glaciers thousands of feet thick. More evidence of their erosive efficiency is displayed in the sets of moraines skirting the eastern edges of the province. Other main features include Hetch Hetchy Valley, the Devil's Postpile, Tuolumne Meadows, Mono Lake, and the Alabama Hills. Major rivers (excluding those mentioned earlier as draining into the Great Valley) include the Walker and Owens rivers.

The Basin Ranges

The Basin Ranges Province is part of the much larger Basin and Range Province of western North America. The Basin Ranges are named after the Great Basin, a region of internal drainage, within which they lie. The dominant feature of the Basin Ranges is sets of near-parallel mountain ranges separated by structural depressions with (usually) internal drainage. The Basin Ranges occur in two areas of California, the northeast and east-central portions of the state. Only the latter is considered here (Day Eight itinerary). Most of the southern Basin Ranges is extremely arid, with large mountain ranges rising abruptly from deep valleys. Most of the structure results from large crustal blocks being lifted (horsts) or dropped (grabens). The lack of vegetation makes the details of the landscape strikingly apparent. The major geomorphic features of the province include the Saline Valley; and the White and Inyo mountains; the Panamint, Last Chance, and Grapevine ranges; and the Funeral and Black mountains. Some have argued for the inclusion of the eastern crest of the Sierra Nevada in this province; inclusion would give it both the highest and lowest locations in the conterminous United States (Mount Whitney and Badwater, respectively). The largest river in this region is the Owens River, now much-reduced by siphoning of water to Los Angeles.

The Mojave Desert

In the strictest sense, the Mojave Desert Province is a sub-unit of the Basin Ranges Province. It has been recognized as distinct in the California context largely because of the region's northern and western delineation by the Garlock and San Andreas faults, respectively. The western reaches of the Mojave include Searles Lake and the Trona Pinnacles, the gold regions of Johannesburg and Randsburg, and the rapidly developing, high-desert urbanization of Palmdale and Lancaster (Days Eight and Nine itineraries). Geomorphologically, the western Mojave is relatively undistinguished, with the major feature being (arguably) the Mojave River. The Mojave River represents the primary drainage of the province, and is an intermittent stream along much of its course from the San Bernardino Mountains across the desert to Soda Lake. It has surface flow only along portions of its channel, except when it floods.

THE WRINKLING EARTH

The Earth's surface is an unstable and dynamic place. The evidence is all around us—earthquakes, volcanoes, fault lines, contorted rocks. All of these features can be explained within the overall framework of plate tectonics, the theory that the outer skin of planet Earth is composed of a series of large, interlocking plates that move relative to each other. The primary force for the movement derives from heat generated within the Earth's interior by radioactive decay. As the rigid surface plates are moved sideways, they might crunch into each other and produce mountain ranges like the Himalaya or Andes. They might break apart with the pieces moving off in opposite directions thereby creating rift val-

Tectonic forces at work: the fractures and displaced layers of rock are evidence of an earthquake. (Lens cap shows scale.) Photograph by Bernard O. Bauer.

Faulting in the LA Basin

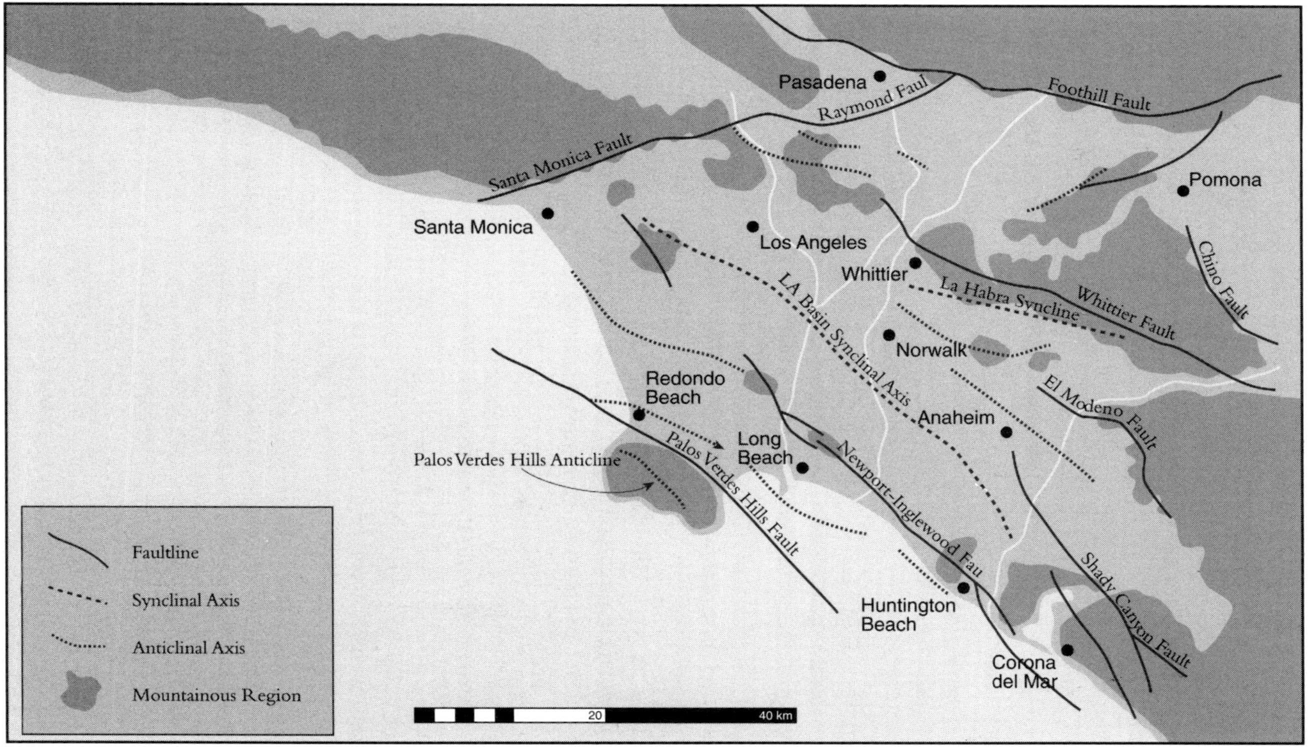
Pasadena
Raymond Faul
Foothill Fault
Santa Monica Fault
Pomona
Santa Monica
Los Angeles
Whittier
La Habra Syncline
Whittier Fault
Chino Fault
LA Basin Synclinal Axis
Norwalk
Redondo Beach
El Modeno Fault
Anaheim
Long Beach
Palos Verdes Hills Anticline
Palos Verdes Hills Fault
Newport-Inglewood Fau
Shady Canyon Fault
Huntington Beach
Corona del Mar
Faultline
Synclinal Axis
Anticlinal Axis
Mountainous Region
20
40 km

leys or new seas like the Red Sea and the Gulf of California, or they might slide past each other creating extensive fault and fracture systems like the San Andreas Fault. The San Andreas Fault marks the boundary between the Pacific Plate, which is moving northwestward at a rate of about 3.4 inches per year, and the North American Plate, which is moving westward at a rate of slightly less than 1 inch per year. This relative movement yields a slow compression, twisting, and fracturing of the existing sediments thereby producing the framework for the topographic features of California.

When a sequence of layered rocks is squeezed or compressed, the layers begin to buckle, much like a rug that is pushed from both sides, creating upwarped regions called anticlines and downwarped regions called synclines. If the compression is continued, the anticlines may peak and fold over.

Most of the broad physiographic features of the landscape take tens of thousands of years to evolve. In most cases, the overall process involves a large number of minute jerks, which are almost imperceptibly small. Nevertheless, when you add up a series of very small movements over many thousands of years the net result can be substantial. In contrast, during extreme events, such as major earthquakes, displacements of five or ten feet occur within less than a minute.

Most of the major earthquake-generating systems in Southern California have their origin in the basement rocks (for example, Foothill, Whittier, Newport-Inglewood, Palos Verdes Hills faults) and many of the features that we see at the surface are related directly to the basement topography. For example, Signal, Dominguez, and Baldwin hills all fall along an extension of the Newport-Inglewood Fault, and similarly, the Puente, Repetto, and Elysian hills follow the trace of the Whittier Fault complex. The Santa Monica–Raymond Fault system is somewhat peculiar because it trends essentially west–east and therefore cuts across the grain of the basement complex. Geologists identify this fault system with the northern terminus of the Peninsular Range geomorphic province, and the fault marks the boundary between the Peninsular and Transverse ranges to the north.

WATER

During the spring of 1991, California was entering the fifth year of a prolonged drought, and government officials and water-supply agencies were beginning to panic. Los Angeles approved a series of water conservation measures that made it illegal to water lawns during the day, to hose down sidewalks and driveways, and to operate decorative fountains with fresh water. Furthermore, mandatory reductions of anywhere from 10 to 25 percent were imposed on residential and commercial water use, with stiff fines levied for non-compliance and plans for cutbacks of up to 50 percent later in the summer. Water supplies for agricultural irrigation were reduced by 75 percent in most parts of the state. Reservoir levels were critically low and there was little hope for relief. In an age of technological wizardry and enlightened management, how could this happen?

There are two dimensions to water resources: supply and demand. Most of California's water falls in the northern part of the state, but most of the population lives in southern California. Many parts of northern California receive more than 50 inches of rainfall annually, whereas the south coast receives only 10 to 15 inches, making it a semiarid environment. Inland deserts such as Death Valley receive less than 2 inches annually and in some years receive no rain at all. An additional complexity stems from the fact that the major topographic features are oriented in a north–south direction. Thus, altitude tends to dominate climatic patterns much more so than latitude even though most of the storm systems that bring rain to California migrate from west to east. The mountains cause air parcels to rise and cool, and, in so doing, initiate a condensation and precipitation cycle. The windward sides of mountains tend to be much wetter than the leeward sides where a "rainshadow" exists, and this explains the lack of precipitation over the inland deserts.

The only reason the Los Angeles metropolitan area can exist in such a dry environment is because of the California State Water Project. This extensive and intricate water-delivery system conveys water from the northern portions of the state to the south via eighteen reservoirs and dams, eight powerplants, seventeen pumping stations, and more than 600 miles of aqueduct—it is the largest state-built water project in the nation. California voters approved a $1.75 billion bond issue in 1960 to begin construction of the infrastructure and the work is still in progress. The major operating costs of the project are paid for by more than thirty water agencies that have long-term water-supply contracts with the state (to the year 2035). The project will ultimately provide for the total delivery of approximately 4.23 million acre-feet.

In addition to the geographical distribution of water supply and demand is the issue of timing. Most of the state's precipitation occurs during November to March and much of it falls as snow, especially in the mountains. Therefore, for a short period during spring snowmelt, a great deal of water flows in the streams and rivers, much of which cannot be used. On the other hand, the demand for water is continuous year-round and peaks during the summer when the growing season is at its climax. Since the summer is the dry season, a great deal of water must be taken from reservoirs or pumped from groundwater aquifers—about 26 percent of U.S. groundwater pumpage occurs in California and most of this is used for agriculture.

The most critical issue with respect to drought, is the substantial variability from year to year in the total volume of available water. Weather patterns fluctuate—average annual runoff for all of California is around 70 million to 75 million acre-feet but this can vary from 18 million acre-feet to 135 million acre-feet. This becomes especially important when there are more than two dry years in succession and there are no intervening wet periods that allow for recovery. Although back-to-back dry years are not uncommon in California, they are becoming increasingly more difficult to deal with because of the growing population and because of the expansion of water-intensive agricultural activities. Extreme droughts of three-year duration are quite uncommon, the last having occurred in 1593 to 1595 (based on tree-ring analysis).

It is interesting to note that the recently mandated water-conservation practices in Los Angeles have caused the price of water to rise in order to offset low consumption rates. Thus, the average household gets less water but pays more per billing period so that the water contractor can maintain the service! Although the economics are sound, the public relations associated with this paradox are rather strained.

PART TWO

The Itinerary

Los Angeles Metropolitan Basin

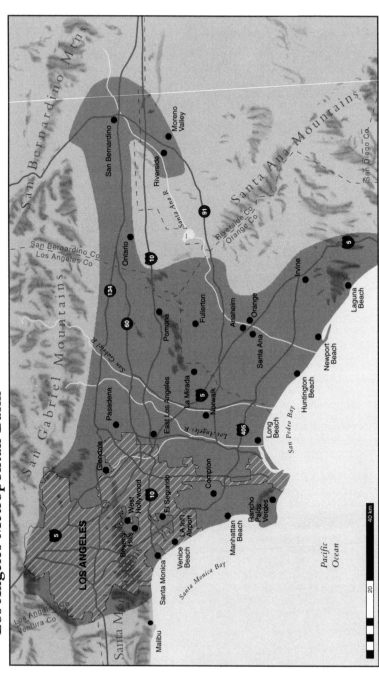

△ *Day One*

LOS ANGELES METROPOLITAN BASIN

Los Angeles is known as L.A. to everyone who lives there (as well as by the vast majority who live in the far western part of the country and most points beyond). This "city" is unlike any other in North America. On the one hand, its huge size differentiates it from all other cities. Picture a city 100 miles wide by 100 miles long. The official city boundaries are rather small—467 square miles—but when residents say they "live in L.A." they may mean anywhere within the entire L.A. "metro" area, which sprawls across a significant segment of the coastal southern California landscape.

The Los Angeles Basin is a large sedimentary system that covers approximately 1,700 square miles with a perimeter open to the ocean on the west but bounded by mountains on the north, east, and south. Topographically the basin can be divided into a northerly and a southerly section. The northerly section encompasses the entire San Gabriel Valley and is bordered by the San Fernando Valley to the northwest (i.e., Sunland/Tujunga area) and by the Chino Basin (i.e., Pomona) to the east. The southerly section is defined by all portions south of a line that traces the axis of the Santa Monica Mountains, Elysian Hills, Repetto Hills, and Puente Hills.

The metro area extends from the northern boundary of Los Angeles County (where it abuts up against Ventura County along the coast and Kern County in the dry interior), and runs 100 miles to the south (including all of Orange County along the coast, the

Freeways encircle Los Angeles's City Hall and Civic Center. Photograph courtesy of the Los Angeles Department of City Planning and Planning.

western portions of Riverside County and the southwestern corner of San Bernardino County in the interior), where the metro area comes in contact with San Diego County and finally ends. In essence, then, living in L.A. can literally mean living in scores of communities situated anywhere along a wide coastal strip that sits north of San Diego County and south of Ventura County, south and southwest of the San Bernardino Mountains, and astride the front and back sides of the San Gabriel Mountains (the latter being Antelope Valley, which is the only portion of L.A. of any notable distance away from the coast).

Because of L.A.'s vast size, residents frequently commute great distances to and from work. It is not uncommon for a Californian to live 50 or 60 miles from work. L.A. lacks a viable public transportation system—the "modern city" dates from the 1880s, from which time many schemes have been put forward to create a viable system, and one even existed for a spell, but none currently exists; however, construction is finally now under way on the first phase of an underground system. Not only must Los Angelenos commute every day by car, but because the recent population has grown much more rapidly than the grid of existing metro-area freeways, the major thoroughfares across the basin are crowded with traffic at most hours of the day. During rush hour, the so-called expressways resemble linear, elevated, tree- or wall-lined nauseating parking lots! With all that idle vehicular traffic spewing exhaust into relatively stagnant air, there is little wonder why L.A. has one of the worst ozone problems found anywhere. L.A.'s tremendous size demands that travelers allot as much time as possible to see the "city."

The route suggested in this guide will only cover a minute parcel of the L.A. metro area, a representative sample of what L.A. is all about.

Walking Tour of Downtown Los Angeles

Begin at the Embassy Hotel, on the northwest corner of Ninth and Grand streets. Built in 1908 and until recently home of the L.A. Philharmonic, the Embassy is now a residential college of the University of Southern California.

Leaving the Embassy, walk north and then west to the southeast corner of Eighth and Hope streets, stop and observe the three residential complexes of what is known as South Park: the Skyline

Los Angeles Walking Tour

1. Embassy Hotel
2. View of South Park Residences
3. Broadway Plaza (Mall)
4. Lobby of Fine Arts Building
5. Overview of 7th St. Market Place
6. Sculpture at Citicorp Plaza
7. Lobby of Post Modern Building
8. Fountain
9. Jonathan Club
10. Lobby of Westin Bonaventure Hotel
11. View of Bunker Hill Residences
12. "The Willows" Garden
13. Wells Fargo Building
14. Trompe l'oeil
15. Central Market
16. Pershing Square
17. Jewelry District

Bunker Hill steps, designed by Lawrence Halprin. Downtown Los Angeles. Photograph by Dennis McClendon.

Condominiums, the Metropolitan Apartments, and the Del Amo housing project. These mark the southern residential area. With the exception of Bunker Hill at the northern edge of downtown, there is almost no other housing available within the city core.

After crossing Eighth and Hope, proceed north along the Broadway building to the entrance on Hope Street, and then through the mall, which is one of several underground shopping areas for the downtown work force. Exit the mall on Seventh Street, the original retail district of L.A., and notice the architecture and building heights, which, due to earthquake hazards, were limited to thirteen stories until 1957.

Now, walk west to Flower Street, cross Seventh Street, and continue west on the north side of Flower Street for half a block to the Fine Arts building, built in 1925 to house the city artists, architects, and drafting teams. The lobby is an engaging mixture of high ceilings, gargoyles, indoor fountain, and tile work.

Continuing west along Seventh Street you will pass the Home Savings building on the corner of Seventh and Figueroa. This is a station stop for the Blue Line, which is the recently completed rapid transit system linking downtown L.A. with the port at San Pedro, to the south. The ceiling mural above the escalators provides a transition from the subterranean environment to the street life.

Crossing both Seventh and Figueroa streets, walk south half a block to the entrance of Seventh Market Place—another subterranean shopping area—with open-air dining two stories down.

Return to Seventh Street and continue west half a block, to the sculpture of the corporate head and the accompanying poem adjacent to the entrance of the Citicorp building.

One-half block farther west, cross Seventh Street and walk up the steps to the lobby of the Coast Savings building—one of the first postmodern buildings built in L.A., completed in 1978.

Exiting the north side of the lobby, stroll east to Figueroa along Wilshire, and then cross Wilshire on the west side of Figueroa to admire the fountain in front of the Swana Bank building.

Continuing north, you will go by the Jonathan Club, site of many of the early transactions affecting the future of L.A.

At Fifth Street cross Figueroa and navigate east, to the skywalk that connects the Arco Plaza with the Westin-Bonaventure Hotel. The lobby and elevators of this futuristic hotel, built in 1978, provide an interesting blend of indoors and outdoors in a science fiction setting.

From the lobby, you should now proceed north on the skywalk over Fourth Street, to the World Trade Center, which is noteworthy for the lack of tenants and the lack of upper stories; it is only three stories high. A short walk north from the lobby, beyond the tennis courts, provides a panorama of Bunker Hill residences, the northern residential core.

Returning to the lobby of the World Trade Center, you will want to walk east, taking the skywalk across Flower Street and into the Security Pacific National Bank, where you will veer right and head south, to the open courtyard known as The Willows. This area of willow trees, grass, and running water provides a peaceful contrast to much of downtown L.A., especially early on Sunday morning.

Newstands in Latino shopping district along Broadway, downtown Los Angeles. Photograph by Dennis McClendon.

From The Willows, backtrack to the main lobby and take the escalators, up and east, to the Hope Street entrance, pass the Calder sculpture, across Hope Street, and directly into the Crocker Center (also known as the Wells Fargo Center). Here are displayed art work by Lawrence Halprin, Jean Dubuffet, Joan Miro, Louise Nevelson, Robert Graham, and Nancy Graves. Take the stairs up and east, and continue east and outside to Grand Avenue. Be sure to note the optical effect of a two-dimensional skyscraper, created by the office tower just a few feet to the north.

Next, proceed south to Fourth Street, taking the escalator down just beyond the overpass, over Fourth Street. Walk east along Fourth Street to Olive Street, where you will be able to admire the trompe l'oeil along the upper stories of the Subway Terminal building, which occupies most of the block.

After walking along the north side of the Terminal building, cross Fourth and Hill Streets and then travel north, to the Grand

Central Public Market. Cross the street to the west of the market, where you will see the Angeles Plaza—the largest housing project for the elderly in the United States. Now, wander east, through the market, enjoying the sights and smells.

Leaving the east entrance of the market, walk south along Broadway, for two blocks, encountering the distinctive sights, sounds, and scents of L.A.'s Hispanic culture.

At Sixth Street, turn right and walk west to Hill Street, where you see a rewarding view of downtown, looking northwest across Pershing Square.

Crossing Hill Street turn left and walk south along the west side of Hill street, as you progress through the jewelry district. At Seventh Street, turn right again and walk west, beyond the L.A. Athletic Club. Then, cross Olive Street to Grand Street, where you will make a right turn and go south two blocks to the Embassy (crossing Grand Street at some point), to be on the west side of the street. You are now back where you began the walking tour.

Driving Tour around L.A.

Our route once again begins within the city proper—near the *University of Southern California*. The university sits in what was once a fashionable, upscale neighborhood. Today, however, the surrounding area is depressed, and gang violence has become a common concern. Once predominately white and middle class, the population of the area is now chiefly black or Hispanic, and distinctly poor. The university, an expensive, private institution, is thus a virtual island of affluence and opportunity amidst a sea of poverty and discontent.

During the daytime, however, the area is both safe and rich with things to do and see: the California Museum of Science and Industry, the famed L.A. Coliseum (seats 105,000 people), Exposition Park (opened 1910), the L.A. County Museum of Natural History, the Sports Arena, and the world famous Shrine Auditorium are all within a short walk of the campus. And don't miss the grandeur of the campus itself, particularly the imposing Bovard

Los Angeles Metropolitan Region

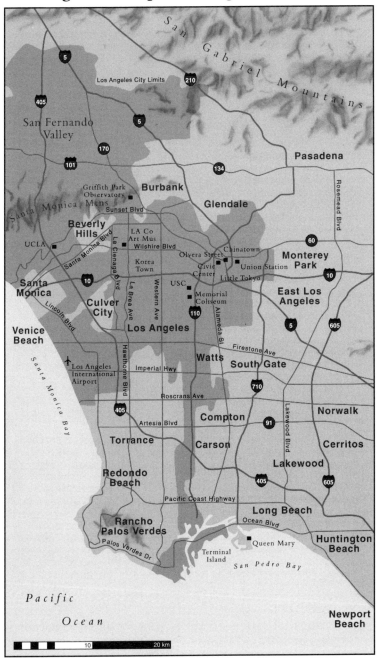

ETHNIC L.A.

Los Angeles is the epitome of an ethnically diverse American city. By some accounts, in fact, there are more ethnic groups in L.A. than in any other American city. The L.A. school system, for instance, alleges that no fewer than 121 distinct languages are spoken by children attending school in the district. The single most commonly spoken language is Spanish—not English.

An ethnically diverse population is good because it indicates that the city, as a whole, is quite cosmopolitan and thus "worldly" in outlook. A cosmopolitan populace is generally open to change; they appreciate differences among peoples of various backgounds; and they are inclined to be generally less provincial, or inward looking, than other populations of the same approximate size. Ethnically diverse, cosmopolitan cities are typically points from which innovative ideas spring and, as such, they tend to be places toward which people wish to move; L.A., of course, is one of the fasting growing cities in the country.

Cities that accommodate a wide assortment of ethnic groups are also great places to live if you enjoy exotic food! When members of a group migrate to a new home they adhere to their traditional dietary habits. This typically means they open restaurants that provide the rest of us with a vast array of tempting delicacies and a brief glimpse into another culture. Mexican food is of course a local favorite here, and there is an abundance of the more traditional, ubiquitous ethnic restaurants—for example, Italian, Chinese, and French among others—but in L.A. the traveler can also feast on uncommon fare, too, such as Korean, Samoan, Guatemalan, Filipino, Haitian.

But there are also less positive characteristics of ethnically diverse cities. When members of a population speak a multi-

Watts Towers, built by the Italian immigrant, Simon Rodia, have become the symbol of a community that is changing from black to Latino. Photograph courtesy of Los Angeles Department of City Planning and Planning.

tude of languages and each group possesses and practices a unique medley of cultural habits, confusion and misunderstanding is almost always guaranteed. Humans tend to malign and denigrate those who are different from themselves. They likewise seem to have less patience for those with whom they cannot communicate readily. As a result, ethnically diverse cities are vulnerable to ethnic conflict.

The gang violence in L.A. can partially be understood in this regard. Each ethnic group has a gang it calls its own. So, there are Korean gangs, Filipino gangs, Vietnamese gangs, black gangs, white gangs, and many other ethnic-based gangs through the L.A. area. The desire, almost the "hunger," to control territory explains some of the dreadful increase in gang warfare across the L.A. metropolitan area.

Will the sorts of difficulties facing L.A. plague other American cities, as they, too, grow and diversify from migration? Or will they learn from this "ethnic test tube" unsympathetically known as the gang capital of America?

Administration Building (Italian Renaissance) and the opposing Edward L. Doheny Jr. Memorial Library building (modified Italian Romanesque). The completely walled, well-patrolled grounds are remnants of L.A.'s yesteryear and a stark reminder of how quickly our inner cities can change.

Leave the University of Southern California area heading north on Figueroa Street. Be sure to notice the large Felix the Cat character atop the Felix Chevrolet dealership; in L.A., it is frequently difficult to tell whether art mirrors reality, or whether it might be the other way around! Proceed to the next major traffic signal, Adams Boulevard, and turn left—but as you do, observe the two large buildings on either side of the road. To your left is the local American Automobile Association headquarters, built in the classic California mission style. To your right is a beautiful example of early twentieth-century Latin American/Catholic (Neo-Churri-

gueresque) architecture, in the form of the 1923 St. Vincent de
Paul Church (which was financed by Edward L. Doheny, a local
oil baron), which is well worth inspection. As you drive west on
Adams, note the old, large buildings that remain from when this
neighborhood was stately and a coveted place to live in the 1930s
and 1940s. Turn right (north) at the next major signal, Hoover
Street, and proceed for a couple of miles along a predominately
Hispanic commercial district.

As you pass Venice Boulevard, look to your right and head
toward Alvarado Street, which branches off on a forty-five degree
angle—but don't proceed on Alvarado. Instead, at that same corner, make a sharp right turn (where you see the handsome Spanish
Seventh-Day Adventist Church) onto Alvarado Terrace and follow
it as it encircles a row of once extremely expensive Victorian
mansions on your left. You can well imagine what the view from
those imposing homes must have been like when there was nothing in front of them for miles.

When you come to Pico Boulevard, turn right, and at the first
major intersection, Union Avenue, turn left (northeast) and observe
that the predominately residential area you just passed through has
immediately converged upon an old run-down commercial area.
The small, swollen humps you drive over are oil "bubbles" or
pockets, each holding valuable pools of petroleum. When you
reach Wilshire Boulevard, turn right and proceed southeast toward
L.A.'s downtown along a portion of the longest central business
district in the world.

Proceed into the downtown area—be alert for alternating one-
way streets—cross over the Harbor Freeway (I-110) and continue
to Grand Avenue (where Wilshire ends) and turn right and then
immediately left onto Seventh Street. Pass the enormous Southern
California Flower Market at Seventh Street and Maple Avenue,
and turn left on San Pedro Street. Two blocks later, turn left onto
Fifth Street (one way), and then turn right onto Los Angeles Street.
This will take you past the main bus station and through L.A.'s
most prominent area for the city's homeless population. Follow
Los Angeles beyond the Civic Center, cross over the Hollywood
Freeway; the road will turn to the right. On your left is Olvera

Street (also known as El Paseo de Los Angeles and possibly L.A.'s first pedestrian mall)—an odd, 1929 re-creation of L.A.'s initial settlement (built with prison labor!) as a Spanish pueblo. Directly ahead is Union Station, a 1939 structure of Mediterranean style; the square clock tower is 135 feet tall.

Los Angeles Street ends here. Turn left (north) onto Alameda Street, and then left again on College, which takes you into Chinatown. Turn left on Broadway—note how paltry L.A.'s Chinatown is by comparison with the population of the city—and then turn right on Sunset Boulevard. Proceed on Sunset until Figueroa Street, turn left and take Figueroa to Third Street, where you must turn right. Make the turn but keep to your left, cross over the Harbor Freeway, and turn left on Beaudry, which you follow to its end (at Sixth Street). Turn right on Sixth Street and then left (two blocks later) on Bixel Street, cross Wilshire Boulevard, and turn right (northwest) onto Seventh Street. Downhill directly ahead is a large, landmark building, the 1938 Bullock's-Wilshire building. Note that you are driving in what used to be an exclusive, expensive hotel district; today it is one of the places where L.A.'s very poor struggle to survive.

HISPANIC L.A.

At Alvarado Street, you enter an inordinately busy, largely Hispanic shopping district—on a Saturday the streets are lined with people from every corner of Latin America. To your right is *MacArthur Park* famous (or notorious) since 1895. In 1991, it was completely fenced and crowned with barbed wire, and the water was drained from the lake to allow for work on L.A.'s proposed underground transit system. Circle the park, observe the people in the park as well as the business establishments (from a large number of Latin American countries) surrounding the park. Note the police subunit on the park grounds, an indication of the excessive crime rate within the park. You are in the very heart of the Rampart Division of the L.A. Police Department, where more murders, robberies, and burglaries occur than in any other part of the city; since 1989, well over 100 homicides per year were recorded within the Rampart Division.

MEXICAN FOOD

Californians enjoy a wide assortment of ethnic foods but, more than any other, they consume Mexican food. Mexican fare is a virtual institution in California, particularly in the southern portion of the state.

Mexican cuisine is distinct from southwestern fare, unlike so-called Tex-Mex, and different as well from Californian food—the latter being a rapidly disappearing native food. It is stout on flavor, mildly spicy, and full of carbohydrates; corn and bean products prevail, as does cheese. Mexican beers—some of which are among the very best in North America—complement most main dishes. They also go well with a fresh bowl of crisp, slightly salted tortilla chips and full-bodied salsa—a Mexican/Californian staple.

The more "Mexican" the restaurant is—that is, the greater the number of employees of Mexican descent working at a particular establishment, and the more customers you find there who are speaking Spanish—the more authentic the food. Chances are you've discovered the real thing. For this reason, it makes no sense to seek out a fashionable, expensive Mexican restaurant in a trendy part of town. There are a multitude of unassuming, reasonably priced Mexican restaurants (as well as small, almost "fast food" stands) located throughout the state (but especially in Southern California) from which to choose. If you find yourself in a predominately Spanish-speaking area, look for a restaurant without an English menu. Remember, as a rule, the worst Mexican food in California (especially Southern California) is assuredly superior to the best you encounter in Idaho, Michigan, and points east. Trust us!

Palm trees in Los Angeles. Photograph by Dennis McClendon.

Less than 3 miles from this point is Dodger Stadium, where the Hispanic population comprises the majority of those in attendance at any given ballgame. After encircling the park and lake, turn right (southwest) back onto Alvarado Street and follow it to Olympic Boulevard, where you turn right. Directly ahead are the three types of palm trees most frequently seen in southern California: the tall ones are California fan palms; the short, scrubby ones are Mexican fan palms; and the wide, exotic looking ones are Canary Island date palms.

As you move along Olympic, a great transformation takes place: The Hispanic neighborhood becomes a Korean neighborhood. By the intersection of Olympic and Vermont Avenue, virtually every sign is in Korean and the transformation is complete. Continue northwest (then west) on Olympic—beyond the Hispanic day-laborers standing on either side of the corner—till you see Western Avenue, where you make another right.

Follow Western to the next major intersection and turn left, which will put you on Wilshire Boulevard again. The imposing green, art deco building on the corner is the Wiltern [Wilshire + Western = Wiltern] Theater, which opened 7 October 1931. Designed by G. Albert Lansburg, the theater was destined for demolition in 1979, when it was rescued by the L.A. Conservancy and kindred concerned parties. After being refurbished, the theater reopened in 1985. Today, this geometric building that sits at what was, in the 1930s, the "busiest intersection in the world," is on the National Register of Historic Places, and has become a landmark within L.A. At this point you are entering another transition zone—where Wilshire will promptly switch from a commercial district to an exclusive residential enclave, only to turn back in to a commercial area by the time you reach Highland (approximately 2 miles west of Western). Note the Scottish Rites Masonic Temple at the corner of Wilshire and Lucerne Boulevard, and one block later look to your left, beyond the old stone gates, as you pass Fremont Place—the street names change to Rossmore Avenue to your left. Fremont Place is an exclusive and extremely private road. Proceed for another six blocks and turn right (north) onto La Brea Avenue.

JEWISH L.A.

You are now driving along what was not too long ago a heavily Jewish district. The side streets to your left and right are still largely inhabited by Jews—who can be seen most readily on Saturday afternoons—but a significant number of "outsiders" are buying into the neighborhood and the variety of stores along La Brea should indicate who they are: relatively young, wealthy, upwardly mobile persons who appreciate and purchase good quality merchandise—those whom the media refers to as yuppies. Observe the synagogues interspersed with stores that cater to the yuppie appetite: "trendy" restaurants, antique purveyors, custom photo labs, expensive clothing stores, art galleries, and so on. When you reach Melrose Avenue, turn left (west) onto a street that is best seen at night.

Hollywood proclaims itself. Photograph courtesy of the Los Angeles Convention and Visitors Bureau.

Melrose Avenue was once a typically mixed-use business area, where any number of goods and services could be purchased. Today, Melrose is perhaps the most trendy street in L.A. and one of the most well-known across the country. Melrose is alive; most stores close only after midnight and some do not close—an interesting phenomenon when you consider the vast Jewish population that lives in the area, since many Jews observe religious customs that forbid such practices. If you are looking for something "wild" and different (especially faddish clothing, from the offbeat hippie boutique to the extremely fashionable), you can probably find it on Melrose between La Brea and Fairfax Avenue. Melrose might come closest to capturing the dichotomous—some would say schizophrenic—psyche that is Southern California. It is worth your time to take a close look and experience the best and worst L.A. has to offer!

After you pass Fairfax High School (on your left), turn left (south) onto Fairfax Avenue. (If you want to see Hollywood, turn right and head north to Sunset Boulevard.) As you drive south on Fairfax, the once completely Jewish area is still evident in many ways between La Brea and Beverly Boulevard. Cantor's restaurant and deli (on your right, at the pedestrian signal between Rosewood and Oakwood avenues) is a long-time landmark and well worth a lunch or dinner stop. Note the historic mural painted on the south side of Fairfax People's Market, just beyond Cantor's. As you cross Beverly, CBS-Television City is on your left, followed by the once-cosmopolitan *Farmer's Market* and the once-trendy Dupar's Restaurant; today, Farmer's Market is perhaps the best example of an L.A. "tourist trap"—note the tour buses that still bring unwary sightseers here to spend their money.

MIRACLE MILE AND BEVERLY HILLS

As you approach Sixth Street, on your left is a bulky black and gold building that extends to Wilshire Boulevard. This is the 1939 *May Company-Miracle Mile* building (under renovation in mid-1991), one of the first department stores constructed with an adjoining parking lot—a department store for commuters. Today, L.A. is full of such stores, most of which are inside shopping malls. The *Miracle Mile* refers to the stretch of Wilshire between La Brea and Fairfax that used to be the symbol of the automobile age in Southern California: buildings that were all designed with the commuter in mind. Beyond Fairfax—moving west into Beverly Hills along Wilshire—you still find a row of automobile dealerships, albeit extremely expensive brands of autos.

Turn right (west) onto Wilshire and head into *Beverly Hills*—a separate city, named after Beverly Farms, Massachusetts, which is not part of L.A.—as you cross San Vicente Boulevard. If you have time, however, first turn left (east) on Wilshire where you will discover the *L.A. County Museum of Art,* the *Page Museum,* and the *La Brea Tar Pits* (where the remains of prehistoric animals continue to emerge). Take note of the sporadic graffiti. This is an inordinately wealthy section of the city, but no part of L.A. is immune from the "outreach" of gang identification, in this case the

WEATHER AND AIR QUALITY IN THE LOS ANGELES BASIN

The physiography of Los Angeles plays a role in the dynamics of local atmospheric conditions, especially air quality. There are essentially two climatic or air-quality seasons in Los Angeles: a hot, dry, smog-filled summer; and a cool, variably-moist, and variably-clear winter. The Pacific Ocean has a strong moderating influence on land temperatures, so coastal locations are substantially cooler than inland valley locations during summer, and relatively moderate during winter. In addition, the maximum daytime and minimum nighttime temperatures are more extreme farther inland. However, the higher mountain regions tend to be wetter and cooler than the lowland valleys.

During the summer, which lasts from mid-April to mid–October, there is little or no precipitation in the Los Angeles Basin and daytime temperatures vary between 70 and 90 degrees Fahrenheit, frequently exceeding 100 degrees Fahrenheit. The driest and hottest period is between July and September when the entire Southwest United States is dominated by an atmospheric high-pressure cell called the Hawaiian High (because that is where it can be found in the winter).

The subsiding air of the Hawaiian High-pressure system also has the effect of capping the Los Angeles Basin through a phenomenon known as a temperature inversion. Thus, all the pollution that is generated at the surface, by industry and more than 10 million cars parked on the freeway, gets trapped—it cannot move inland because of the ring of mountains, it cannot move upward because of the inversion, and it cannot move out over the ocean because the prevailing winds are onshore. In fact, the prevailing wind patterns create an air-quality gradient from the coastal zone, which characteristically has the best quality air, to the inland areas of the San Gabriel Valley (Pasadena, Pomona, Riverside, and San Bernardino), where we generally find the worst quality air. On average, the L.A. Basin experiences approximately 160 days of unhealthful air quality (based on ozone concentrations) each year, and almost all of these occur during the summer. Fortunately, the Hawaiian High-pressure system begins to weaken and move offshore around October. This spells relief for the Basin because temperatures drop and air quality improves. The area comes under the influence of the westerly winds, which bring sequences of low-pressure, cyclonic storms. Thus, the weather in the winter switches between two- to five-day cold, wet, overcast periods, and one- to fourteen-day cool, clear periods. Air quality improves because the upper-atmospheric cap no longer exists, allowing pollutants to rise above the mountains, and because the rain tends to cleanse the air of particulate matter. Occasionally, a Santa Ana condition arises, whereby the air flow reverses direction, taking pollutants offshore. Daytime temperatures during the winter average between 40 and 60 degrees Fahrenheit with minima dipping into the low thirties. Most of the annual rainfall of about 15 inches (although the total varies considerably from year to year) is received in December to March, and much of this falls as snow in the mountains at elevations above 5,000 feet.

distinguishing inscriptions that mark ethnic territories. Businesses in this area now have to budget to barricade the facades of their buildings or plan for the unending scrubbing and repainting of walls.

The elegant *Wilshire Theater,* followed by a statue of John Wayne, shortly on your left signals the entrance to what once was Beverly Hills's renowned theater district. Many of the theaters have since been replaced or modified for other purposes. Just beyond the Wilshire Theater is La Cienega Boulevard. If you have

Smog and the Inversion Layer in the LA Basin

time, turn right and visit the remnants of L.A.'s acclaimed "restaurant row." Continue driving west along Wilshire. In addition to the remarkable display of extravagant automobiles, be sure to notice the equally vast array of domestic and international banks (notably Asian and Middle Eastern) on this stretch of the boulevard; indeed, very few banks licensed for business in Southern California do not have a branch (often their main office) on this portion of L.A.'s famed linear downtown.

As you drive west, glance or drive down some of the narrow side streets, especially the residential ones. Note the well-manicured lawns and parkways (owned by the city) lining every roadway, but note, also, the rather small homes and lots, as well as the number (and type) of cars in each driveway. The well-maintained, picturesque stucco houses date from the 1930s and 1940s, when typical houses were small by today's standards. Likewise, in the 1930s, cars were a luxury and few residents owned more than one. Today, however, almost everyone owns or has access to more than one car, a necessity in a sprawling metropolis that lacks adequate public transportation.

On your left you will see the Beverly Pavilion Hotel at the intersection of Wilshire and Crescent Drive. A block later, at Canon Drive, you begin to pass through the heart of Beverly Hills' suave central business district. The Beverly Theatre is on your left; also on your left is the extraordinarily extravagant Beverly Regent Hotel, on Wilshire between El Camino and Rodeo. Now you are about to enter one of the most swank, fashionable districts in North America: Next comes Beverly Drive, followed immediately by the celebrated Rodeo Drive. It is worth driving up and down these (and the ensuing two or three) streets to get a sense of the lavish decadence exhibited in so few blocks. Be sure to see the stylized Italian piazza on the corner of Rodeo and Dayton Way.

Follow Wilshire past the noteworthy intersection with the two Santa Monica Boulevards, where you find Wilson's House of Leather (a popular store during the 1970s) and the Beverly Hilton Hotel, both on your left; the dense cluster of high-rise buildings to your left is Century City. The first Santa Monica Boulevard you cross is so-called Little Santa Monica. This stretch exists almost exclusively in Beverly Hills, and was fashioned as the commercial, pedestrian portion of Santa Monica Boulevard in this area; the larger, "real" Santa Monica Boulevard is almost always busy with vehicular traffic and runs alongside Beverly Hills' "green belt." A short distance after you pass the hotel, you will pass the L.A. Country Club and reenter the city of L.A. and Wilshire becomes entirely residential for several miles.

You are approaching Westwood, home of the *University of California at Los Angeles* (UCLA). Along this stretch of Wilshire is the only row of high-rise condominiums in Los Angeles that compares favorably with those in Manhattan. The Bunker Hill portion of the downtown area contains high-rises, but not to the same extent. In fact, this might be the only other part of the country where the residents prefer (and can afford) a very expensive high-rise lifestyle. As in Manhattan, the cost of the condos increases in direct proportion to the floor number—the higher the number, the higher the price tag. Virtually every building in this area was once a low-rise hotel or apartment. When Wilshire reverts back to a commercial strip, you have entered bustling Westwood. On weekend evenings, Westwood Village (to your right) is always crowded with young people (many from UCLA, also to your right). This, combined with the very limited parking facilities, makes it a difficult but interesting area to visit.

MARINA DEL REY AND VENICE

As you cross Veteran Avenue, note an imposing, rectangular, white structure on your left. This is a local federal building, which, because of its proximity to UCLA, is frequently the scene of civil demonstrations. Across the street is the Veteran's Cemetery (one of the largest) and Administration Building. Straight ahead is I-405, the San Diego Freeway. Take I-405 southbound; the on-ramp is on the far side of the freeway, on the right. Stay on I-405 for approximately 5 miles. You will first cross the intersection with I-10—for many years the busiest, and still one of the busiest, intersections in the world—after which the Marina Expressway (Highway 90) will come into view. Those are the Baldwin Hills on your left, which are anticlinal folds and full of oil; note the derricks on top. Still on

I-405, you go in and out of Culver City, and take Highway 90 west, in the direction of *Marina Del Rey*. Driving west on 90, Loyola Marymount University is visible on your left, atop another oil-laden anticline.

Take 90 to its terminus—at which point it becomes the Marina Expressway—and follow it to Mindanao Way and turn left. Cross Lincoln Boulevard (California 1), and enter the city of Marina Del Rey. Turn right on Admiralty Way and follow this road around the perimeter of this artificial marina. Note the high density of yachts and moderately expensive restaurants along Admiralty. As the road begins to veer to the left, you will see a large duck pond on your right. This is part of an almost linear, slightly unsightly wildlife sanctuary, and, as might be imagined, yacht owners have exerted considerable pressure to purge the area to harmonize with their fastidious slips. Pass the Marina International Hotel, and turn right at Via Marina (moving quickly to your left) and then make a left onto Washington Street (and return to the city of L.A.–Venice Beach). Follow Washington till it ends (about eight blocks), and you will find yourself on the Venice Fishing Pier. On weekends, it may be difficult to park close to the pier; this area is frequented by roller skaters, tourists, fisherpersons, residents out for a walk, and a large cross-section of Southern Californians.

When you leave the pier, drive north along Speedway (the slender, westernmost road that connects with the pier entrance) for several blocks, before turning right. Drive all the way to South Venice Boulevard (about 0.5 mile) and turn right. Along this road you will see how drastically densely settled this portion of *Venice* is. The homes are small—some are tiny—but quite expensive nonetheless. This area was settled some time ago, and has become a coveted place to live. Turn right onto Pacific Avenue, which you will follow south until it ends.

As you cross Washington Street you will notice how much larger the lot sizes here are. A short distance farther you will see a canal on your left and a long row of truly expensive, customized homes behind it, up on the rise, connected by walking bridges. The canal is a remnant of the once-extensive Venice canal system. The concrete waterways were built in the early part of the century by

Abbott Kinney, and at one time they extended a full 15 miles. Venice was modeled after its namesake in Italy, but the concept never quite caught on. That row of homes beyond the canal is relatively new, but many hearken back to Kinney's dream of Italian Renaissance mansions lining the canal route—the homes average about 5,000 square feet, and many of these one-of-a-kind houses sell for one million dollars or more!

As you drive along Pacific Avenue you are on the backside of the Marina, and shortly you come to a "L" in the road where you can park and capture an excellent view of the marina's entrance. Turn left onto the unmarked road that parallels the harbor. This eventually becomes Via Marina; turn right when you reach the signal at Admiralty Way, and follow this road until it ends (at Fiji Way, beyond Mindanao). Turn right, toward Fisherman's Village (open daily , 10:00 a.m. to 9:00 p.m. in winter, until 10:00 p.m. in summer). This road is a dead end. As you make the loop, the area to your right is vacant land, another protected refuge and a part of Ballona Wetlands. Follow the road back to Lincoln Boulevard (California 1) and turn right.

Drive south on Lincoln Boulevard for several miles. You will pass a good deal of industry along the way, most of which is aircraft and airline–related. Eventually you cross Manchester Avenue and enter the community of Westchester, and almost immediately thereafter Lincoln merges with and becomes Sepulveda Boulevard; follow California 1, to be sure. At this point you drive through a tunnel, beneath one of the main runways at Los Angeles International Airport (LAX), and when you emerge on the other side you cross Imperial Highway and the new Century Freeway (estimated completion date, 1993).

EL SEGUNDO, L.A. HARBOR, AND LONG BEACH HARBOR

You are now in the city of *El Segundo*. *Segundo* means second in Spanish, and this city was so named by the Standard Oil Company in 1911 to signify its second refinery in California. El Segundo has been a predominantly industrial and manufacturing center ever since. When you reach El Segundo Boulevard (where you will see a large Hughes Air building), turn left and head toward I-405 (San

Diego Freeway), taking note of the aircraft-related businesses along the way. Take I-405 south till you reach I-110 (Harbor Freeway) south, which you should follow to the L.A. Harbor in San Pedro. Along the way, you pass through mostly industrial areas—such as Hawthorne (named after the novelist), Redondo Beach, Torrance, Carson, and Wilmington (originally New San Pedro). A significant number of oil wells and refineries come into view as you approach the harbor; relicts from an era when oil was king in the L.A. Basin.

Exit I-110 at Harbor Boulevard (it is also the exit to the Vincent Thomas Bridge, which you pass on your way to the off-ramp) and stay to your right as the road curves under the freeway. Follow Harbor (south) toward Ports O' Call. Turn left at Sixth Street and enjoy an afternoon at the L.A. Maritime Museum (straight ahead and open Tuesday through Sunday, 10:00 a.m. to 5:00 p.m.) and the Ports O' Call Village (to your right and open daily, 11:00 a.m. to 9:00 p.m.). Inexpensive harbor tours can be taken at Berth 77. Fisherman's Wharf, at the south end of Ports O' Call Village, is home to San Pedro's fishing fleet.

The *Los Angeles Harbor* (discovered in 1542 by Juan Rodríguez Cabrillo) has officially changed its name to Worldport L.A. It is now the largest human-constructed harbor in the western hemisphere, and some of the largest commercial and passenger ships in the world dock here. It has become the busiest harbor in the country, as well as the busiest passenger port on the West Coast and thus serviced by all major Pacific Ocean cruise-ship lines—including convenient service to Catalina Island. And it is home of the famed *Love Boat*. Across the channel from Fisherman's Wharf, at the entrance to the harbor, is Terminal Island and two noteworthy institutions—the renowned U.S. federal prison and the select L.A. Yacht Club!

Leave Ports O' Call via Harbor Boulevard (north) in the direction of the Vincent Thomas Bridge. Harbor changes its name several times as it winds its way around the container yards of the harbor; it becomes Front Street, then Pacific Avenue, then John S. Gibson Boulevard, and finally B Street, which you will follow to Anaheim Street. Turn right on Anaheim, and take this road to I-710

The Big Curl—Los Angeles's tribute to surfing. Photograph by Paul F. Starrs.

(Long Beach Freeway) south. This will take you to the Long Beach Harbor (as you observe the signs that direct you to the port). The *Long Beach Harbor*—most total cargo tonnage on the West Coast—is home to the *Queen Mary* (the largest luxury liner afloat) and the *Spruce Goose* (the 200-ton wooden airplane built by Howard Hughes), which is on exhibit inside the world's largest aluminum geodesic dome (both are open daily, 10:00 a.m. to 6:00 p.m.; they stay open later in spring and summer).

If you have time, drive along Harbor Scenic Drive (beyond the *Spruce Goose*) till it ends (at Pier J, toward Berth 249), and pull into the public fishing area on the left. From this point you will find a good, wide-angle vista of the downtown Long Beach Marina, Seal Beach (farther south along the coast), and then Sunset, Huntington, and Newport beaches (beyond Seal Beach), all of which are popular Southern California surfing areas. The tall structures off-shore are oil derricks.

When you depart the harbor area, head back toward I-710 but look for Panorama Drive and turn right. This will take you to Queensway Drive, which is the scenic alternative and well worth the negligible extra time it takes. Queensway offers the traveler an attractive, ground-level perspective of downtown Long Beach, as well as numerous parking and picnic areas. Continue until Queensway ends, at which point it becomes I-710 (north). Take I-710 (north) till it merges with I-5 (a twenty-minute drive), and be sure to take I-5 northbound.

Follow I-5 north to the Los Feliz Boulevard off-ramp (another twenty-minute drive), turn left (under I-5) onto Los Feliz toward Hollywood, and look for signs that direct you to the Griffith Park Observatory (part of Griffith Park, a 4,000-acre municipal enclave owned and operated by the City of Los Angeles). On the opposite side of the park is the Gene Autry Western Heritage Museum, which affords the traveler an opportunity to relive the "wild west" (closed Mondays; phone for hours, 213-667-2000). Remain on Los Feliz till you arrive at Hillhurst Avenue and turn right. Hillhurst will become Vermont Avenue at the stop sign, and you should stay on Vermont and proceed uphill into the park. Follow the signs to the observatory, past the Greek Theatre, and park your car. The Griffith Park Observatory—named after Colonel Griffith J. Griffith, who donated the park, the theater, and the observatory to the city—has something for all ages, and one should allow at least ninety minutes to appreciate both the view (when possible) and what awaits inside. The observatory, built in 1935, contains a twelve-inch Zeiss telescope. More than 1.5 million people enjoy it each year and it is well worth a stop (closed Mondays; phone for hours, 213-664-1191).

△ *Day Two*

ACROSS THE LITTLE DIVIDE—
LOS ANGELES TO MORRO BAY

Today we cross four counties (Los Angeles, Ventura, Santa
Barbara, and San Luis Obispo) and three geomorphic provinces
(Peninsular Ranges, Transverse Ranges, Coast Ranges). The
entire region is underlain primarily by sedimentary rocks of
varying thickness that have been faulted, folded, tilted, warped,
twisted, and in many locations intruded by volcanic rocks from
deep within the Earth's interior—the geology is complex, to say
the least. Most of the sedimentary rocks are geologically relatively young (less than about 60 million years) and originated in
marine environments of varying water depth. These rocks have
been substantially uplifted and little eroded—a good indication
of the recency of mountain building. Lateral compression has
played a major role in shaping the region, giving rise to the
three major lowland areas (Los Angeles Basin, Ventura Basin,
Santa Maria Basin) and the intervening mountain ranges (Santa
Monica/San Gabriel, Santa Ynez/San Rafael). These tectonic
processes continue today, albeit intermittently, as evidenced by
the frequency of earthquakes all along the West Coast of the
United States.

Los Angeles County lies at the northern terminus of the Peninsular Range. This system of mountains extends from south of the
Mexican border to the northwest almost parallel to the coast. It
differs somewhat from the others in being mostly igneous and
metamorphic rocks. A major intrusion of molten lava in the form

Los Angeles to Morro Bay

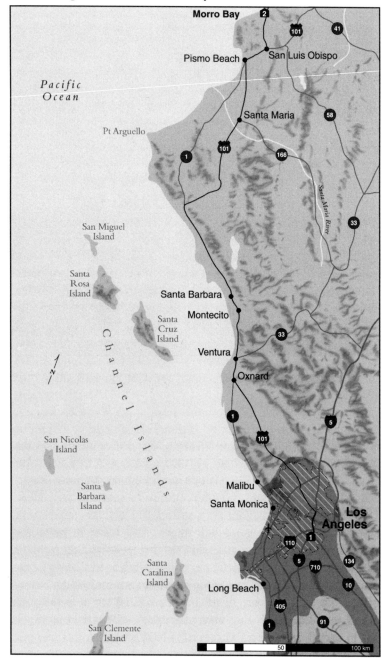

Morro Bay

2

101

41

Pismo Beach

San Luis Obispo

Pacific
Ocean

Santa Maria

58

Pt Arguello

1

101

166

Santa Maria River

33

San Miguel
Island

Santa
Rosa
Island

Santa Barbara

Montecito

Santa
Cruz
Island

33

Ventura

Oxnard

N

Channel Islands

1

San Nicolas
Island

5

101

Santa
Barbara
Island

Malibu

Santa Monica

Los
Angeles

1

San
Catalina
Island

110

5

710

134

10

Long Beach

405

91

San Clemente
Island

1

50   100 km

of a huge batholith occurred approximately 90 million to 100 million years ago in the southwestern portion of the state, and these rocks have effectively resisted the constant erosive processes leaving behind a steep-sided but gentle-topped topography. The system actually extends westward out into the Pacific Ocean—the southern Channel Islands (Santa Catalina, San Clemente, San Nicolas, Santa Barbara) are simply the exposed remnants of upland areas.

The mountain ranges immediately to the north of the Los Angeles metropolitan area, including the San Bernardino, San Gabriel, Santa Susana, and Santa Ynez mountains, are part of the Transverse Range geomorphic province. The name "transverse" derives from the east-west orientation of the axis of these mountains, which is counter to the southeast. Thus, the Transverse Ranges cut across the grain dividing the southern and central portions of the state. This is no small matter since some of the peaks are the highest in the state reaching elevations greater than 10,000 feet—a wonderful situation for skiers, but a formidable barrier to surface modes of transportation. In fact, the only places where road and railway corridors have penetrated the mountains are where the mountains have been faulted by tectonic activity, such as at Cajon Pass or at Fort Tejon. The Transverse Ranges extend from the Mojave Desert–Salton Trough area in the eastern interior of the state, through Point Arguello on the west coast, and into the Pacific Ocean—there is no natural break at the coast. Evidence of this extension into the ocean is manifested by the Channel Islands off Santa Barbara, which are aligned along an extension of the Santa Monica Mountains known as the ancient Cabrillo Peninsula. Scientists have recently discovered a major tectonic boundary under the ocean in alignment with the Transverse Ranges known as the Murray Fracture Zone, and the two are thought to be related.

To the north of the Transverse Ranges is the Coast Range geomorphic province which trends north-south. This system of coastal mountains and intervening valleys extends from the Santa Ynez River in the south to beyond the California-Oregon border in the north.

The Tour

MALIBU TO POINT MUGU

Begin at the intersection of the Pacific Coast Highway (or Highway 1) and Malibu Canyon Road by Pepperdine University. The traveler can take numerous routes from downtown Los Angeles. The main route follows Highway 10 westbound from central Los Angeles and merges with the Pacific Coast Highway in Santa Monica. Northbound on the Pacific Coast Highway (Highway 1) will take you along the coast past the Pacific Palisades, the Getty Museum, Topanga Beach, and directly into Malibu.

Malibu is of particular interest to those of you who are familiar with the many "beach" movies produced in the 1960s. The image is one of sunshine, surfing, volleyball, and tanned teenagers partying on wide stretches of beach—remember Annette and Frankie? In reality, and especially today, this narrow strip of beach is largely

Hindu temple, Malibu Canyon. Photograph by Elliot McIntire.

inaccessible to the public. As you drive north along the Pacific Coast Highway past Malibu (actually, you will be driving westward even though this is a north-south route) you will see steep-sided cliffs on the one side of the road and an essentially solid wall of houses and fences on the beach side—only occasionally will you glimpse the beach. Many of these houses were at one time or another owned by people in the movie industry and used as weekend beach houses. Although many are still used only on weekends, they have been remodeled and expanded, and most movie stars can't afford them! The wealthy landowners are very protective of "their" beach frontage and there has been considerable controversy about access to these beaches. Fortunately, private ownership extends only to the high-water line, and some public access pathways lead from the road to the beach. However, there is essentially no parking along this stretch of the highway, and chronic beach erosion by storm waves usually means that the water line is directly underneath some of these stilted houses. During an intense storm in January 1988, waves invaded the basements of some of these houses and bulldozers were used during low tide to build sand berms in front of the houses to prevent a recurrence.

Paradise Cove is approximately 5 miles north of Pepperdine University. This small embayment is typical of many small, natural havens for fishing and recreational craft along the southern coast of California. Its main claim to fame, however, is the parking lot in which James Garner's trailer was parked for the TV series "The Rockford Files." Paradise Cove is situated on the sheltered side of *Point Dume,* one of the many large promontories that juts out into the Pacific Ocean. These points usually exist because they have a core of very resistant rock that is not easily eroded by the elements, whereas the adjacent sediments are gradually worn back. Ultimately, the point may become separated from the mainland to form a stack or mini island.

Immediately to the north of Point Dume are a series of lengthy and wide beaches (Zuma, Trancas, Nicholas Canyon, Leo Carrillo). These are either county or state beaches that remain undeveloped and there is lots of parking and sand, even on crowded

PROTECTING THE SHORELINE

At many points along the coastline, you will see heavy investment in a variety of engineering techniques to protect the shoreline. So many of California's resources—recreation, shipping, industry, defense, real estate, and highways—depend on a stable shoreline and inviting beaches that the prevention of coastal erosion and flooding has become big business here.

The wide beaches at Santa Monica Bay, for example, are the result of beach nourishment. Millions of cubic yards of sand are brought in by barge, truck, or pipeline and dumped on the beach (over the last fifty years, more than 500,000 cubic yards per year along Santa Monica Bay to stretch the beaches by several hundred feet). Since more than 100 million visitors come to California beaches each year—60 million in Los Angeles County alone—the beaches' contribution to local economies justifies the expense of $5 to $10 per cubic yard for beach nourishment and preservation.

Groin (also spelled groyne) structures are built perpendicular to the beach with the shoreward end usually established inland of the high-water line. Groins are frequently incorrectly referred to as jetties. The groin reduces the alongshore transport of sand by acting as a dam. However, this action also results in some erosion immediately downdrift of the structure because of the interrupted sediment supply. Good examples of this effect can be seen in the groin fields at Ventura.

Although most groins in California are built of stone (or "rip-rap") they may also be constructed of wood, steel, or concrete. Groins are among the least expensive means of providing local relief from beach erosion. However, the deleterious downdrift erosion frequently makes their use impractical.

Seawalls are common shore-protection structures along the southern California coast. Seawalls, unlike beach nourishment and groin projects, are designed to protect resources behind the beach. Many California seawalls are relatively unobtrusive, with much of their bulk hidden by the enhanced sediment volumes of nourished beaches.

Examples of seawalls can be seen at several locations along our route, including Santa Monica, along the coast road north of Ventura, and at Santa Barbara. Seawalls and groins are considered hard protection, as opposed to the soft protection afforded by beach nourishment and coastal sand dunes. More than 130 miles of the California coast contain some hard protection, an increase of more than 400 percent since 1971. These figures include about 75 percent of the coastline between Ventura and Carpenteria and a similar amount for northern Monterey Bay.

Aids to navigation and harbor improvement are also important structural elements of the California coast. They are of two main types: *jetties* and *breakwaters*. Jetties are designed to stabilize entrances to harbors. Their design also focuses tidal currents, enhancing jetlike flows that sweep sediments from the channel mouth. Large stone jetties can be seen at the entrances to Marina del Rey, Port Hueneme, Ventura Harbor, and Morro Bay.

Breakwaters are used to reduce or eliminate the effects of waves at inlets, or to enclose areas to provide calm anchorages. Examples of the former are associated with the jetties at Marina del Rey and Port Hueneme, among others. Closer views of jetties and this type of breakwater are conveniently obtained near Morro Rock.

Much of the California coast is relatively straight, providing only poor natural anchorages. Several major breakwater projects have been built to improve the quality of harbor and port facilities. The largest of these is the breakwater enclosing the ports of Los Angeles and Long Beach.

A smaller breakwater encloses the harbor at Santa Barbara. Begun in 1927, this project has been controversial because of its location and apparent effects on local beaches. By blocking alongshore sediment transport toward the south, the breakwater caused more than 1,000 feet of accretion at Leadbetter Beach on its northern side. Concomitant large-scale erosion on beaches immediately south of the structure is also attributed to the harbor works. Decades-long litigation has sought to fix responsibility and liability for erosion as far south of the harbor as Carpenteria.

days. Public services are maintained, and in certain sections overnight trailer parking is allowed. As you proceed north, look for patterns of pronounced sand accumulation. Leo Carrillo State Beach marks the boundary between Los Angeles and Ventura counties.

You can park south of *Point Mugu,* but almost all spaces are on the ocean side of the road so be careful of oncoming traffic. This is a recommended stop for many reasons. In the forefront

POINT DUME

If you have time, take a side trip to Point Dume and hike to the top (approximately five minutes along well-marked trails). The view is spectacular and as you look back to the mainland toward the south you can see a good example of an eroding marine terrace and coastal bluff system. A plaque briefly describes the history of the area. There has been increased pressure lately to develop the land on Point Dume, but fortunately it has met with strong local resistance.

is a gravel beach with only a thin apron of sand which gradually becomes wider and wider toward the south. In the far south, you see a huge sand sheet that appears to have crossed the road and climbed the slope of the mountain to heights of almost 200 feet. This feature formed because the wind transported sand from the beach across the road, and up the base of the mountains.

As you scan the horizon offshore, you may be able to see some of the Channel Islands. Unfortunately, this stretch of the coast is often foggy in the early mornings and your visibility may be limited. If this is the case, then you might want to take a short walk around the ocean side of the point to get a sense of the awesome power of waves to erode rock. You will find the remnants of the old Pacific Coast Highway, which, undermined by wave erosion, collapsed. Be careful, however, because big waves can splash you and the rocks are often wet and treacherous.

Adventuresome and able-bodied travelers may want to get closer to the water to look at the rocks. The rocks contain evidence of a previous world—a marine world of fine-grained sediments and burrowing animals. These rocks originated in the Tertiary (around 25 million years ago) and contain fossils of scallop-like animals and of tube-shaped tunnels formed by burrowing shrimp.

As you proceed north from Point Mugu, the underlying rocks get progressively younger until you reach the Santa Clara River, and then they get progressively older again. The reason is that the Ventura Basin is another bowl-shaped structure (syncline), like the Los Angeles Basin, which was formed through compression of the many layers of rocks from the sides. The Point Mugu region and the Santa Monica Mountains represent the southern lip of the bowl, the Santa Clara River region represents the bottom of the bowl, and the Santa Ynez Mountains form the northern lip. The bottom of the bowl has been filled with Tertiary and Quaternary (less than 2 million years) sediments, just as in the Los Angeles Basin, and the process continues today. In some places, the sediments are as thick as 20,000 feet, which makes it one of the thickest sequences in the world. The result of this infilling is the broad Oxnard Plain.

OXNARD PLAIN TO CHANNEL ISLANDS NATIONAL PARK

The *Oxnard Plain* is a rich and productive agricultural region that produces many garden crops and flowers (for seeds) for local and national consumption. Many of the fields have a series of propeller towers that are turned on during periods of frost. The idea is that cooler air forming close to the ground overnight is mixed with warmer air slightly higher in the atmosphere thereby preventing damage to the crops. In addition to agriculture, the main sources of income for this region are petroleum, tourism, and the naval base at Port Hueneme. These rather diverse activities have caused many problems, both physical and human. For example, to maintain productive levels of agriculture, groundwater aquifers have been pumped intensively. This, along with pumping oil from the ground, has caused large-scale subsidence of the land in some places. Furthermore, the lowering of the water table to levels below the sea has caused an invasion of sea water into the sediments. Such saline and brackish waters are not suitable for agricultural or residential uses, and usually the damage is irreversible.

The human dimension is often more complex. Much of the agricultural activity is supported by cheap, often undocumented labor and has given rise to a large Hispanic population that is migrant and linked to the growing season. The military personnel are also somewhat migratory but at cycles linked to tours of duty. They are dominantly young, white or black males. In contrast, the tourist industry centers around recreational activities along the coast, and this industry caters to a wealthy, non-resident population, dominated by white families and "yuppies." This mix of economic and cultural backgrounds gives rise to considerable tension.

From Point Mugu follow the Pacific Coast Highway north past Port Hueneme and exit at Channel Islands Boulevard westbound. The latter road traverses a cross-section of cultural space including many small shops, banks, the naval base with private golf course, and eventually an expensive residential area. The residential area consists of large houses with automobile access in the front and boat access in the back via a complex system of roads and water canals. Turn right (north) on Harbor Boulevard for about three blocks to Oceanaire Street. Follow Oceanaire Street to Mandalay Beach Road and turn right on Mandalay Beach Road.

This stretch of beach is known as the *Oxnard Shores;* you will be struck by the width of the beach and the abundance of sand. Mandalay Beach Road ends at West Fifth Street, which is also the southern boundary of Mandalay State Beach. There, a system of small dunes backs the beach. Closest to the beach are the active foredunes and farther inland the dunes get progressively older and more stable because of the dense vegetation. Most of the sand came from the Santa Clara River, which empties into the Pacific Ocean a few miles to the north. The housing district you just drove through used to look exactly like this nature park. Note that the residences are inundated with blowing sand (look at the wind-breaks across the doorways to the houses), and the owners must keep a constant vigil against the accumulation of sand on the roads and on their property—a small, but never-ending tax for locating on the coast in spite of nature's best warnings.

Follow West Fifth Street for about 1 mile and turn left (north) on Harbor Boulevard. On the left you will see Mandalay Generating Station and a series of large oil-storage tanks in the middle of a marshy wetland—a curious combination to say the least! A little farther north, you cross the Santa Clara River. In drought years, it is little more than a trickle, but this belies its immense power to transport sediment during wet periods. If time allows, you may want to visit the *Channel Islands National Park Visitor's Center.* Turn left (west) on Spinaker Drive. The center is at the end of Spinaker Drive, and it houses a series of displays that characterize the geology, geography, history, and biology of the Channel Islands. Of particular interest are the tidepool petting display and the spiral staircase walking tour of the various levels of the ocean. As a reward, there is a telescope on the top view deck (only three stories high) that can be used to view the islands and harbor area. There are also a small book and map shop and clean restrooms. Walk to the beach and examine the many different types of engineering structures built to stabilize the shore and protect the harbor—this is a good example of a heavily engi-neered coastline.

CHANNEL ISLANDS

The eight islands in the coastal waters of Southern California are divided into two groups according to their association with landform assemblages identified on the mainland. The southern group—*San Clemente, Santa Catalina, San Nicolas,* and *Santa Barbara*—is related to the Peninsular Range geomorphic province, whereas the second group—*Anacapa, Santa Cruz, Santa Rosa,* and *San Miguel*—makes up the northern group, which is an offshore extension of the Santa Monica Mountains, part of the Transverse Range geomorphic province. The northern group is thought to have been connected to the mainland a few million years ago in the form of a continuous strip of land known as the Cabrillo Peninsula.

Santa Barbara Channel was at that time a bay open to the ocean at its western margin. This allowed continental species of animals and plants to migrate to and inhabit the peninsular region. Today, there are many species of animals and plants on the islands that are similar to those found on the mainland. However, because of tectonic activity and sea-level fluctuations, the old peninsula has been gradually covered by water thereby isolating the high ground as islands. Thus, there are many endemic species of plants and animals (including the island fox, the scrub jay, and more than sixty species of rare or endangered plants) that differ considerably from their mainland counterparts because of extensive periods of isolation. Fossils of pygmy elephants or mammoths have been found on Santa Rosa and San Miguel, and some controversy has developed over their origins. One faction contends that these animals migrated overland when the peninsula existed, while other suggest that a few mammoths swam from the mainland and reproduced and evolved on the islands.

Similar arguments surround the origins of many plant species whose seeds might have floated across the ocean or could have been transported by birds. The islands serve as protected rookeries and nesting grounds for eleven species of seabirds and six species of pinnipeds (aquatic mammals with finlike flippers including seals and walruses). The marine ecosystem is extraordinarily diverse and productive because of the mixing of cool and warm waters in this region.

Despite their proximity to the mainland, the islands are relatively inaccessible to the public and are among the most sparsely populated parts of this region. Only Santa Catalina and Anacapa have ferry service, although recreational boaters are allowed mooring on some of the other islands. Their recreational and economic utility is limited by private, military, and government ownership; by environmental restrictions; by rugged topography and lack of building space; and by limited boat landings and navigation hazards. Santa Catalina, for example, was almost entirely owned by the Wrigley family (of chewing gum fame) until recently, except for Avalon, which is still the only incorporated town on any of the islands. Most of the islands are managed by the National Park Service or by The Nature Conservancy—Santa Catalina is now managed, in part, by the Catalina Conservancy. Santa Barbara and Anacapa Islands were declared national monuments in 1938, and in 1980, by an act of the U.S. Congress, the Channel Islands National Park was established to include San Miguel, Santa Rosa, Santa Cruz, Anacapa, and Santa Barbara Islands and all waters extending 1 nautical mile around them. Long-term management is critical to the delicate environment of these islands because of the lack of fresh-water supplies and because of a limited carrying capacity for humans and wildlife. For example, the introduction of sheep, goats, and bison to some of these islands has resulted in overgrazing and removal of the grasses and shrubs, which in turn leads to landslides and soil erosion.

VENTURA

Whether you decide to stop at the Visitor's Center or not, you will eventually follow Harbor Boulevard north until South Seaward Avenue where you turn right and proceed over Highway 101. South Seaward Avenue continues through a residential district of southern *Ventura*; you are driving up a gentle slope that marks the flank of a marine terrace. Turn left on Main Street and note the small-town mission architecture lining the streets. Proceed approximately 2 miles and turn right on North California Street and you will be facing the San Buenaventura City Hall (the official name for the Ventura City Hall) built in 1912 to serve as the Ventura County Courthouse.

The road veers to the right around a statue of Junipero Serra (Father of the California missions); turn left on Poli Street in front of the City Hall. Continue on Poli Street and note the Victorian houses on the right. The road begins to veer to the right. Make a sharp right turn on Ferro Street and proceed up the very steep hill to the scenic overlook at Serra's Cross in Grant Park. The original cross was moved here from the mission, although the cross you see now is a replacement of the original, which burned shortly after the turn of the century.

To the south are the beaches of Ventura, Oxnard, and Port Hueneme—note the many groins emplaced along the shore to maintain the beaches in a reasonably sandy state. On a clear day, you can see the large breakwater protecting Port Hueneme. Directly in front of you is the Ventura pier, which used to serve as an oil-loading dock but is now used primarily for recreation. To the north, on your right, is the Ventura River valley and as your eye scans from the coast toward the coastal hills you will see a staircase-like profile that shows clearly a sequence of marine terraces. Note how the city of Ventura below is shaped like an "L," following the contours of the coast and then bending east up the Ventura River valley.

Proceed back down the hill, but bear left at the first junction below the lookout, on Summit Drive, following the scenic drive signs. Turn right on Kalorama. This lane leads you on an alternate route back to Poli Street. Turn right, return to North California

Street and turn left. Turn right at the next intersection onto Main Street. This section of downtown Ventura has been extensively "boutiqued," with a variety of antique shops, clothing stores, and restaurants. Main Street will bring you to the reconstructed mission. Mission San Buenaventura was the ninth of the California missions, and the last founded by Father Serra. Relicts of an ancient Chumash Indian site have been recovered at an excavation adjacent to the mission. Dates for the campsite range back as far as about 1500 BC. Continue north along Main Street, until it rejoins Highway 101 on the northern extremes of the city.

Follow Highway 101 north to *Emma Wood State Beach and County Park.* The park is on a thin strip of land protected from the waves by large, stone-rubble rip-rap. During high-tide storms, large waves overtop the sea wall and often toss large rocks onto the road. This road used to be the old Pacific Coast Highway, which was relocated higher and farther inland to its present position for safety and to lessen maintenance problems. The new Pacific Coast Highway along this stretch of coast up to Santa Barbara is built on a series of alluvial plains and marine terraces. These topographic features are easily identified by their relatively flat or gently sloping surfaces, which are set against the steep, rugged slopes of the older mountains. Marine terraces are formed through deposition of sediments in the coastal ocean by streams and landslides (note the many landslide scars), reworking of the sediments by waves and currents to form a smooth, flat table-like, submarine feature, and subsequent uplifting of this feature out of the sea by tectonic forces. Some of the best examples of well-developed terraces occur just to the west of the Ventura River, where five successive terraces are preserved (other good examples are at Palos Verdes Peninsula and San Clemente Island). The remains of mastodon and prehistoric horses have been found in these terrace sediments. Commonly, the broad table-like features are dissected by rivers, streams, and debris flows that originate in the mountain canyons and cut into the softer coastal sediments during wet periods.

If you look to the south along the coast, you can see waves breaking a considerable distance offshore. Shoaling waves such as these indicate that the water is very shallow. The Ventura River has

COASTAL CURRENTS

The coast of California is dominated by a gentle flow of cold water known as the *California Current*. This current origi- nates in the Gulf of Alaska and meanders southward along the west coast of North America to the tip of Baja California where it turns westward into the open equatorial ocean.

During spring and summer (March to September) there is substantial "upwelling" along the coast as surface waters are driven offshore by winds and cool waters from great depths rise to replace the void. A by-product of this upwelling is the transport of nutrients from the lower layers to the upper, light-dominated ("photic") layers; this greatly enhances ma- rine productivity and activity. During early fall (September and October) the wind patterns shift and the upwelling ceases; coastal waters are at their warmest making for pleasant recre- ation, but poorer fishing. During winter (November through February), the cold waters of the California Current are forced farther offshore.

Major disruptions to these patterns occur about every six to eight years because of the *El Niño* ("Christ Child") phe- nomenon, when winter atmospheric pressure and wind systems change their usual orientation and reverse their di- rection. As a result, warm equatorial waters begin to flow from the west Pacific toward Central America where the current splits into two arms, one moving north and the other south. In the vicinity of California, this means that the normal cool surface waters, offshore flows, and upwelling, become instead "abnormally" warm surface water, onshore flow, and downwelling. Marine productivity is adversely affected by the warm water, and during some extreme El Niño events, irreparable damage to marine ecosystems and fisheries has resulted. El Niño events also yield abnormal increases in rainfall along coastal mountains, creating landslide and debris-flow hazards.

The shape of the coast and the topography of the continental shelf also influence the pattern of coastal currents. The classic example occurs within the Santa Barbara Channel. As the California Current migrates southward past Point Conception, the sharp east–west trend of coastline creates a large eddy system that brings a warm current of water close to shore moving from Los Angeles, past Santa Barbara, toward Point Conception. The zone of mixing of warm and cool waters near Point Conception creates ideal conditions for a broad range of marine organisms, so that both northern-water species and southern-water species can be found in proximity. The Channel Islands also create unusual wave conditions along the shoreline of the mainland because waves can only pass through the gaps between the islands. Thus, many locations on shore are sheltered from the effects of extreme storm waves even though neighboring locations are battered periodically because they are directly in line with the wave gap. Oil companies, taking advantage of this natural protection, have located many of their offshore oil platforms in the wave shadow-zone in the lee of the islands.

deposited a great deal of sediment in the ocean and a submarine delta has formed. Bathymetric maps that show underwater contours for 1855 and 1933 indicate that approximately 580 feet of beach had accumulated in the vicinity of Ventura during this seventy-eight year period. As waves travel toward shore, they encounter this shallow delta deposit and begin to change their shape—they transform from deep-water to shallow-water waves that have steep forward faces and foamy crests. The delta may eventually be uplifted out of the sea and form a marine terrace similar to the ones upon which the Pacific Coast Highway is constructed.

PUNTA GORDA TO MONTECITO

As you approach *Punta Gorda* on your way north, keep your eye on the upper sections of the rock exposures on the landward side of the highway. Just before you reach Punta Gorda you will see an example of a recent marine terrace deposit (Pleistocene) that sits on top of much older and tilted sediments (Pliocene). There is an "age gap" between these groups of sediments and the layers of the marine sediments above the age gap cut across or, in geologic parlance, rest unconformably on the steeply dipping layers underneath. This indicates that a period of intense tectonic activity and intense erosion occurred before the terrace deposits on top were laid down—another geological detective story solved.

The stretch of coast from *Sea Cliff* to Santa Barbara will offer the traveler a first-hand view of many aspects of oil production along coasts. The mountainsides are laced with pipelines, and "herds of iron horses" indicate the location of subterranean pools of oil and natural gas. At Sea Cliff, an extensive system of piers extends over the ocean to allow vehicular traffic to a series of offshore wells. On a clear day, you can see a huge number of offshore oil platforms and derricks in deeper water between the Channel Islands and the mainland. Just to the north is a small community called Mussel Shoals and there you can see a small island with an oil well on it. This island may look peculiar despite the palm trees, because it is an artificial island built with large stone rubble. Its only reason for existence is the oil that lies beneath it.

At *La Conchita* (2 to 3 miles north of Sea Cliff on the right), you will find a biogeographical interest story and potential stop for banana connoisseurs. Exit at La Conchita (right-hand exit) and turn left at the stop sign. Drive to the end of the road (about 1 mile). There you will find a small, commercial banana plantation that raises more than sixty varieties of bananas, including exotic species such as Manzano with its distinctive apple fragrance, and Blue Java, which boasts a tangy, citrus-like flavor. Most horticulturalists would claim that banana trees rarely bear fruit in the Southern California climate. But the La Conchita soils and microclimatic conditions appear to be suitable for growing high-quality

OIL PRODUCTION IN CALIFORNIA

In 1987 California produced 400 million barrels of oil a year from 242 oil fields (over 51,000 wells) in seventeen counties. It ranks fourth in oil production by state and contains almost 20 percent of the United States's proven reserves. Commercial production began in 1876, yielding 22 billion barrels of crude oil in the years since then.

Oil and natural gas form when plant or animal matter are buried in sediment and preserved from decay. Gradually the debris is transformed by a slow series of chemical reactions into liquid oil and natural gas. Where large pools of water, oil, and gas are trapped in permeable rock under a layer of impermeable rock, oil drilling is economically feasible. Most of the oil in California has come from rocks in the Pliocene and Miocene strata, formed 2 to 26 million years ago.

Where the pool is not trapped, the petroleum can seep upward to the earth's surface to create tar pits and other easily identifiable traces of the oil below. The earliest oil prospectors in California sought out these oil seeps in accessible coastal areas in the Los Angeles Basin, Santa Cruz, and Half Moon Bay, for instance. Californian crude oil was transported by truck, pipe, and, as early as 1880, by oil tankers.

New understanding of the geological structures that created oil fields and the landforms overlying them allowed a four-fold increase in California oil discovery and production in the first third of the twentieth century. Still more sophisticated techniques of surveying and drilling have enabled oil companies to exploit oil deposits in sedimentary basins very deep under the ground (many wells in the state go down 10,000 feet): Los Angeles Basin, Ventura Basin, Santa Maria Basin, San Joaquin Valley. Kern County's wells account for nearly two-thirds of the state's production.

Oil Production

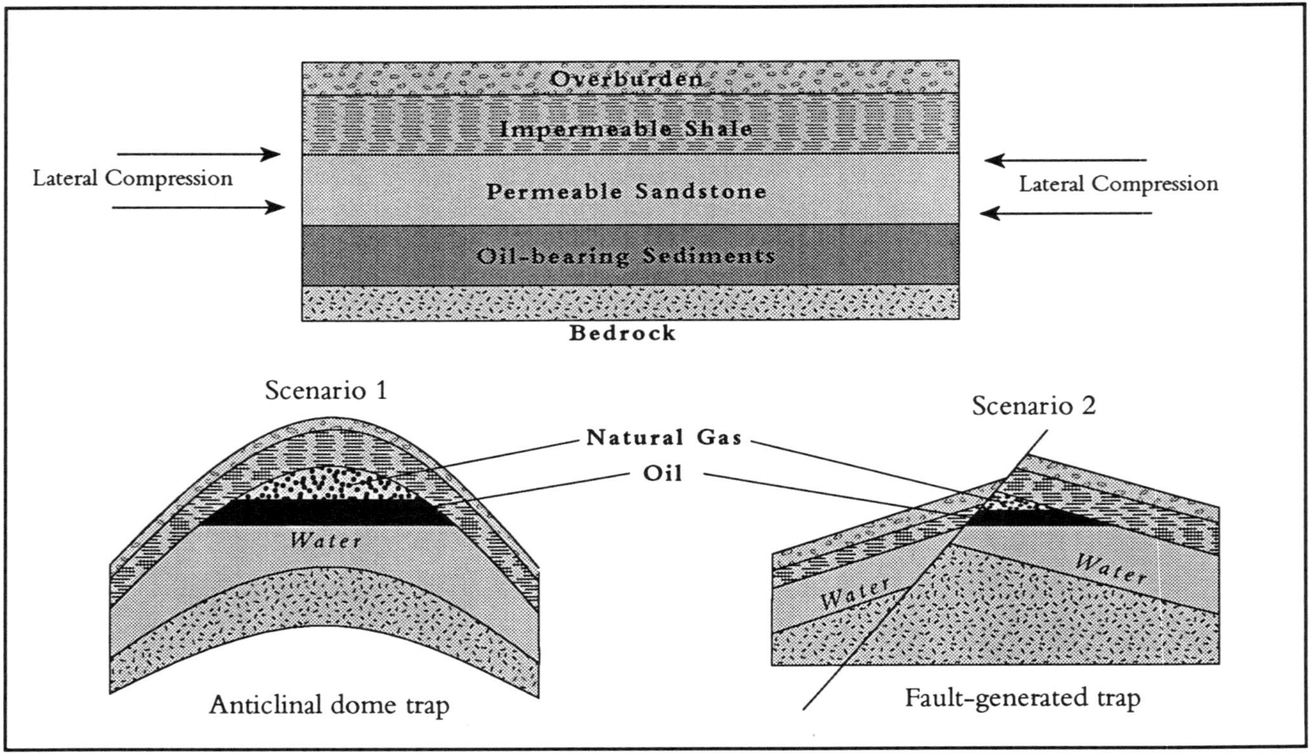

Overburden

Impermeable Shale

Permeable Sandstone

Oil-bearing Sediments

Bedrock

Lateral Compression

Lateral Compression

Lateral Compression

Natural Gas

Oil

Scenario 1

Scenario 2

Water

Water

Water

Water

Anticlinal dome trap

Fault-generated trap

California Oil and Gas Fields

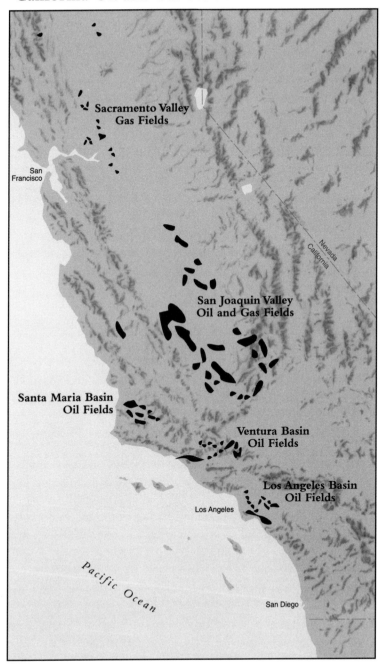

Sacramento Valley
Gas Fields

San
Francisco

Nevada
California

San Joaquin Valley
Oil and Gas Fields

Santa Maria Basin
Oil Fields

Ventura Basin
Oil Fields

Los Angeles Basin
Oil Fields

Los Angeles

Pacific Ocean

San Diego

California's oil reserves are being depleted rapidly even as production is decreasing. The only largely unexplored region for new discoveries is off shore (offshore fields currently account for 15 percent of the state's production). Californians are acutely aware of major environmental disasters caused by oil spills in the ocean—the 1970 Santa Barbara oil spill, the *Exxon Valdez* grounding in Alaska, the deliberately destructive "spill" into the coastal waters of Kuwait—and there is strong political opposition to allowing new oil drilling offshore. Moreover, tough Environmental Protection Agency guidelines on air pollution in Los Angeles and other large cities are stimulating interest in alternative sources of energy. Unless oil companies manage to improve their refining processes so that emission levels are reduced, it seems likely that the demand for California's oil and gas will continue to decline.

banana fruit. The banana bunches are covered in polyethylene bags to prevent damage from frost and to accelerate ripening. The fruit is sold to the public as it becomes ripe, if you care to experiment.

Approximately two thirds of the way from Punta Gorda to Rincon Point is a sea cliff that shows an overturned section of rocks (older rocks on top of younger rocks) and an exposed view of the Rincon Fault. At the top is the Pleistocene terrace and unconformity referred to previously. Directly beneath is a large, bare area with a distinctive reddish brown color due to burning. The fuel for the fire is not the vegetation, but the oil-saturated shale. Near the top is a strong odor of burning oil and sulfur. Historically, this vent was used as a landmark by seafarers because white smoke and steam often rise from the rock, especially after rainy periods. The stretch of coast by Rincon Point boasts as many as ten marine terraces. Can you pick them out? The highest is on the order of about 1,300 feet above sea level. As you pass by Rincon Point, you cross the border from Ventura County to Santa Barbara County.

When the early Spanish explorers traveled the coast of California, they found Indians in the *Carpinteria Basin* who were particularly skilled at building wooden boats made waterproof with the local tar. The explorers named this place "Carpinteria" meaning carpenter shop. The tar was taken from the tar ("brea") pits to the south of the town (present site of the city dump), where fossil remains of Pleistocene vertebrates have been found. The tar is derived from oil seeping upward form the underlying, oil-bearing shales of much older age—the volatiles within the petroleum mixture evaporate through time leaving a tarry substance behind. Such upward seepage of oil occurs in many places both onshore and offshore, and contributes in no small part to the considerable oil-pollution problems typical of this stretch of coast. Human-related accidents are not the only source of oil pollution.

Summerland is just a short distance north of Carpinteria and it is central to the Summerland Oil Field, one of the oldest in the state and the first to be developed in the ocean. Oil production began in 1887 and the number of producing wells on piers ballooned to more than 300 by 1900. Surprisingly, oil was never found in the deeper Miocene shales; rather, it accumulated in the upper Pleistocene sediments. In the late 1920s, production began to slow down, and the field was eventually closed. All that remains today are a few scattered pilings along the coast.

Proceed north on Highway 101, take the Olive Mill Road exit, and proceed straight across the intersection onto Coast Village Road, which takes you into Montecito's tourist and business area. *Montecito* is an old millionaire's enclave that was originally the home of most of the area's wealth before Santa Barbara's growth— note that Montecito is still a separate, incorporated city. Many successful industrialists and notable celebrities including John Steinbeck, Winston Churchill, and the Kennedys have lived or retreated here for extensive periods. Coast Village Road eventually curves under Highway 101 and continues along the shoreline of Santa Barbara as Cabrillo Boulevard. Follow Cabrillo for approximately 2 miles and note the extensive well-groomed beach and adjoining park with outdoor volleyball facilities. This beach is heavily used by locals and visitors alike—especially those who choose to stay right across the street at Fess Parker's Hotel. (Remember him as Walt Disney's Davy Crockett?)

SANTA BARBARA

As you approach the pier you will note the Santa Barbara Visitor's Center on your right. Turn right immediately after the center onto Santa Barbara Street and follow it past Highway 101. Approximately 2 to 3 miles past 101 is Mission Street where you should turn right and then left on Laguna Street, which takes you into the mission parking area. This is the "Queen" of California missions, and you should allow at least an hour to investigate this part of the state's distinctive heritage. You'll see the mission's influence on the architecture of many buildings in Santa Barbara.

Upon leaving the Mission grounds, turn right on Los Olivos Road and turn left on Castillo Street, which leads back down to the waterfront and pier. As you approach the waterfront, keep your eyes open for parking spaces. These can be particularly scarce on weekends and during the summer when there are many community-sponsored events. A walk down the pier (officially known as Stearns Wharf) offers both sight-seeing and lunch. Three fine restaurants specialize in seafood and all have window seating for those who are willing to wait. Alternatively, there is a seafood take-out stand close to the end of the pier where you can get a quick shrimp cocktail or steaming-hot bowl of chowder; or, if you have a bit more time, you can select your own lobster or crab, which will be cooked for you as you wait. Be aware however that the local waterfowl (gulls and pelicans) pester you for handouts (and perhaps entertain you) as you sit and snack at the picnic tables.

On your way back down the pier you can stop at one of the small curio shops to browse or at the ice-cream parlor for dessert. As you make your way down the pier, be sure to look at the character of the beach on either side—the extensive deposit of sand on the north side of the breakwater is known as Leadbetter Beach. It owes its existence entirely to sedimentation patterns induced through construction of the breakwater in 1929 to provide a harbor for fishing and recreational boats. As a result, the beach to the south

Santa Barbara City Hall shows the strong influence of the Spanish colonial architecture. Photograph by Dennis McClendon.

Manzanita tree, with Santa Barbara in the distance. Photograph by Robert A. Rundstrom.

has been starved of sediments and is therefore much narrower than it once was.

Follow Cabrillo Boulevard north along the shoreline skirting Leadbetter. Cabrillo turns into Shoreline Drive, which eventually veers right at La Mesa Park and turns into Meigs Road, which, in turn, becomes Carrillo Street. From the traveler's point of view it is just one continuous strip of asphalt, so you can ignore the change in names. Eventually you reach Highway 101 where you should proceed northbound. After two or three exits, get off again at the State Street Ramp (it is also labeled Highway 154 in the direction of Cachuma Lake). Highway 154 north will take you up the steep, coastal face of the Santa Ynez Mountains, part of the Transverse Ranges. As you wind your way up toward San Marcos Pass, watch out for signs of a spectacular brushfire that occurred in 1990; it devastated the local hillsides and burned down many expensive homes. Although plants soon covered the charred hill-sides again, it will be a long time before the original second- and third-generation species will again dominate. You may still see bales of hay damming smaller creeks and incisions—these were emplaced right after the fire to prevent winter rains stripping the soil off the hillsides and delivering the sediment onto the road. However, the extraordinarily dry weather that allowed the fire to spread so fast also saved the area from massive soil erosion.

Immediately after the summit, turn left on Old Stage Coach Road. This windy route was once traveled by stagecoaches enroute from the east toward the coast. The Cold Spring Tavern was built to serve the coaches and their passengers, and it provides a scenic, cool oasis in the otherwise dry, hot mountain terrain. It is worth a stop, if not for a cool beer, then just to examine the paraphernalia from a time long gone. The tavern was recently "discovered" by weekend biker gangs composed of Santa Barbara professionals, by and large—that is, accountants, doctors, lawyers, and anyone else who prefers to dress in black biking leather and can afford expensive motorcycles as weekend "freedom machines." On a Sunday afternoon you will have to park a considerable distance along the road, and walk through the forest canopy past the BMWs and Mercedes to enjoy the rustic, no-frills environment.

Old Stage Coach Road winds its way underneath the Cold Spring Arch Bridge and comes to a stop. Turn left at the stop sign, and proceed north on Highway 154. Cachuma Lake is an artificial reservoir that serves as one of the main water supplies for Goleta and Santa Barbara. It was built in 1953 and it backs up the Santa Ynez River. The lake has been stocked with trout from Kamloops, British Columbia, and the surrounding campground and recreational facilities are in great demand. When there is a drought, however, the lake levels will be down considerably, making it difficult to use boat-launching and docking facilities.

SANTA YNEZ TO CASMALIA

Turn left (west) on Highway 246 and watch for your first winery. In the town of *Santa Ynez*, turn right on Meadowvale, then left on Sagunto, the main street. Sagunto turns into Questa, which takes you back to Highway 246. The town's personality is split between agriculture and tourism: it's a fake western village, an affluent retiree community, and home for commuters to Santa Barbara. As you get back on Highway 246, heading west to Solvang, you will see a bingo parlor, a venture of the Santa Inez Indian Reservation (a Chumash Indian reservation)—bingo for profit is legal only on reservation lands. Cross Refugio Road (which leads to Reagan's ranch). Mission Santa Inez (the old spelling) will be on your left as you come into Solvang.

Plan to stop in *Solvang* to try Danish pastries. In 1911 Danish immigrants founded a folk-school here, and the town kept its character as an authentic Danish village for nearly half a century. In the 1950s and 1960s it became cute and touristy.

Continue west on Highway 246, crossing Highway 101 at Buellton. Stop in Buellton to sample Andersen's split pea soup. Admire the flower fields in Lompoc Valley, the source of almost 50 percent of the world's flower seeds. As you enter *Lompoc*, you'll pass La Purisma Concepcion Mission State Park. Many historians feel that La Purisma is the most authentic and beautiful of all California missions. In Lompoc, Highway 246 crosses Highway 1. Turn right (north) on Highway 1, and then turn left (northwest) onto Lompoc-Casmalia Road. The road passes through part of the huge Vandenberg Air Force Base, Lompoc's biggest employer. Here you are moving from the Transverse Range geomorphic province to the Coast Range province. Drive through Casmalia, winding through ravines, terraces, and agricultural land, and rejoin Highway 1, going north to Guadalupe. In Guadalupe, turn left onto West Main Street (Highway 166) and go west to the coast and Rancho Guadalupe Dunes County Park (no admission charge).

SANTA MARIA AND GUADALUPE

The towns of *Santa Maria* and *Guadalupe* are within the Santa Maria Valley, which is another large synclinal structure filled with alluvium (similar to the Oxnard Plain and Los Angeles Basin) and used extensively for agriculture. The county park is on the southern portion of a large littoral cell bounded by Point Sal on the south and Point San Luis on the north. Thus, large accumulations of

LITTORAL CELLS

What determines where beaches develop? It is useful to think of the California coast as series of isolated "littoral cells." A littoral cell usually has one dominant source of sand and other sediments, and one dominant sink of sediments, with a transport system of currents and winds carrying the sand from source to sink. The source might be a river or an extensive bluff or headland system, whereas the sink might be a submarine canyon, a sand-mining operation, or another headland. Most littoral cells along the coast of Southern California have a scallop or comma shape consisting of a northerly source and a southerly sink. Since the prevailing alongshore currents move sediment southward, sediments usually accumulate along the southern portions of these scallops. Most major headlands (Point Mugu, Point Dume, Point Arguello, Point Sal, for example) have extensive beaches on their northern flanks where the current slows down, whereas their southern flanks are rocky and barren because the current has been directed offshore and what little sediment remains in transport is deposited offshore. If the downdrift sink of a littoral cell is a submarine canyon, the sediments move into the deep ocean through submarine landslides and are lost from the nearshore system. Ordinarily this means that there is no beach deposit, and the sand cannot by recycled into a neighboring littoral cell. Such losses of sediment usually lead to downdrift erosion, and similar situations arise through intensive sand mining or through the construction of extensively engineered coastal-protection structures designed to trap sediments locally.

In the past, the dominant source of sediments to the coast has been rivers and streams. These are the conveyor belts that move ground-up rocks from the coastal mountains and uplands to the lowland basins and nearshore systems. In so doing, they grind the rocks up further, round them off, and sort them by size and weight (the smallest and lightest particles tend to move the greatest distances). However, over the past thirty to forty years most rivers and streams have been tamed through the construction of large dams (more than 1,200 dams in California alone) that provide flood protection, generate electrical power and supply water. These dams also act as large settling basins trapping all but the finest of sediments being transported downstream. More than 50 percent of the sand supply has been cut off by damming rivers, and the beaches of California have undergone substantial erosion since the construction of these dams. Only in northern California is there a constant supply of sediment to the nearshore because there was no need to dam the streams and rivers in the early days, and they are now protected by the Wild and Scenic Rivers Act of 1972.

sediment occur in the form of wide beaches and large sand dunes. As you drive into the park you will see a small sand-mining operation. The sand on the landward side of this extensive dune system is well-sorted by the wind that carries it this far inland—so much so that very little sieving and processing is required for commercial use of the dominantly quartz sand in the production of glass and as aggregate.

As you proceed toward the beach, be aware that the land is leased by the county for grazing as a source of revenue—you are advised to proceed slowly, lest you have a close encounter with a bovine. The multi-use philosophy of these county parks becomes even more apparent at the beach, where you will find beach-goers picnicking and playing on the beach within a stone's toss of pumping oil wells. Note the extensive dune system to the south (Mussel Rock Dunes) dotted with numerous small, erosive mounds of sand or "hillocks," which are stabilized by the clumps of vegetation growing on them. Much of the sand close to the beach is very mobile, but underneath the mobile sand sheet are older stable dunes. In fact, Mussel Rock, which acts as a promontory in this littoral cell, is an ancient dune that stands 450 feet high and sits on an older sedimentary rock formation. The park is a transitional zone between northern and southern plant communities. Typically sand verbena, sea rocket, and beach morning glory grow here in the shifting sands, whereas mock heather and dune lupine grow farther inland.

At the mouth of Santa Maria River is a 365-acre wetland that serves as a breeding-ground and habitat for many species of water-fowl, as well as snakes and larger animals, including mule deer, weasel, and gray fox. The underlying sedimentary rocks contain fossils of large birds and whales. The extensive beach and dune system owes its existence, in part, to the Santa Maria River. The river supplies a significant volume of sediment to the coastal littoral cell through erosion of the mountain and valley rocks. This erosive power of the river can be seen in two ways. First, there are magnificent examples of river terraces and cut banks along the margins of the river valley, especially around the town of Santa Maria. Second, the county line that separates Santa Barbara and San Luis Obispo counties generally follows the course of the river but snakes its way back and forth across it. In fact, the county line follows the course of the river as it must have been quite some time ago, indicating that the river has shifted its position through time. Indeed, the shifting of river courses, except for engineered aqueducts, is a very common natural process. Such changes have been the basis for political tension and intense border-property disputes. Retrace your route back out of the park toward Guadalupe and proceed north on Highway 1.

DUNE LAKES TO PISMO BEACH

Follow Highway 1 north past the Guadalupe dunes on your far left and continue past Oso Flaco Lake Road. Oso Flaco Lake is a small filled-in depression fed by Oso Flaco Creek before it drains into the ocean. The lake received its name from soldiers of the Portola expedition who camped there and were fortunate enough to feed themselves on a large grizzly bear they shot (*oso flaco* means lean bear in Spanish). Highway 1 continues north through a series of turns that skirt the edge of The Nipomo Mesa, a gently undulating to level terrace that has been incised to produce the Arroyo Grande and Santa Maria Valleys. A series of small lakes, called the *Dune Lakes,* are visible to the west. In essence, these lakes are simply interdune depressions (eroded low-lying scallops sculpted by wind flowing over and through blowouts in the foredune system) that have become waterlogged through groundwater seepage from the mesa. They formed approximately 15,000 years ago and they are the only coastal freshwater lakes remaining in California. Unfortunately, increasing pressures from urban development and long-term groundwater pumping has essentially cut off the source of fresh water, and the lakes have become stagnant as a result of agricultural waste-water runoff. Black Lake, the most southerly of the series, is so named because of its dark color, which is in part due to underlying peat deposits. The northernmost lake (Big Pocket), on the other hand, is now dry. All are privately owned.

As you proceed north on Highway 1 (locally called the Cabrillo Highway), you negotiate a series of turns and eventually reach Halcyon Road and turn left. Halcyon Road drops abruptly down the face of the terrace on which you are driving onto an agricultural plain (Arroyo Grande Valley). The vista is spectacular. Turn left onto Cienage Road, which takes you past the Callender Dunes on your way to Oceano. Collectively, the Mussel Rock Dunes, Guadalupe Dunes, and Callender Dunes are known as the Nipomo Dunes. They cover 18 square miles and are some of the least developed dune systems in California. The oldest remnants of this extensive dune sheet can be found on the Nipomo Mesa, where they extend back approximately 18,000 years.

As you travel through Oceano keep a look out for a distinctive old house on the righthand side that is surrounded by house-trailers. It is the "Coffee Tea Rice House" built in 1885. *Oceano* was once a major shipping and distribution center for the Arroyo Grande Valley, whereas now it serves primarily as a service center for the thousands of tourists that flock here every year to enjoy the seaside. Cienage Road turns into Cabrillo Road, which you take past Grover City and into Pismo Beach. A short walk out to the

Expensive condominiums replace natural sand dunes at Pismo Beach.
Photograph by Bernard O. Bauer.

Pismo Beach Pier shows you the quintessential California beach
life-style—wide beaches, surfers, neon beach-wear, boardwalk
eateries, video parlors, and plenty of sunshine.

It is worth looking to the south for an oceanfront panorama of
the extensive dune fields that you have driven past for the last two
hours. Immediately to the north are Avila Bay and San Luis Bay.
The pier was built in 1881 and has undergone very little renovation
or reparation. Its original purpose was for loading locally grown
produce and unloading agricultural supplies. However, even as
early as 1895 it was given over to recreational purposes when a
dance pavilion was constructed at its foot (the pavilion only lasted
until 1920), and today it serves as a tourist landmark and sport-
fishing facility. Indeed, Pismo Beach was one of the earliest California recreation areas, one that became known as a summer
getaway for the well-to-do. The trend continues today, as witnessed by the extensive, luxurious development taking place
everywhere. For the less-than-well-to-do, there are many campgrounds and recreational-vehicle parks to take advantage of.

Pismo Beach was originally called Pismo, which is derived from the Chumash Indian word for asphalt tar seeps. The tar was used for waterproofing canoes and baskets, as was the case up and down the coast of California. However, the town is best known for the Pismo clam, which was so abundant in the past that it was used as animal feed. The clams were harvested through a simple plowing process which was so effective that a limit of 200 per day was imposed as early as 1911. By the mid-1940s the commercial harvest had depleted the beds so that only sport-clamming was allowed, and now the sport limit is restricted to ten clams per day. It is quite common to see people "raking" the sands for these delightful morsels; however, tourists should inform themselves of the local regulations on tainted shellfish. Pismo Beach is one of the only remaining places in California where it is still legal to drive on the beach.

PISMO BEACH TO MORRO BAY

Our route out of Pismo Beach continues north on Highway 101, past Shell Beach and then turns inland to cross the San Luis Mountain Range. The coastline from Avila Beach to Morro Bay is largely inaccessible because of the rugged, steep topography. Indeed, most of the peninsula is undeveloped and there are only a few small dirt roads with restricted access. The region close to Point Buchon was part of Rancho Canada de los Osos y Pecho y Islay, which translates as "the valley of bears, bravery, and wild cherry"—perhaps better left for a more ambitious journey. The other place of note along this stretch is Diablo Canyon, which received more than its fair share of news coverage because of the construction of the Diablo Canyon Nuclear Power Plant (1968 to 1978). Part of the controversy stemmed from opposition to all kinds of nuclear-power generation, but most of the controversy stems from the plant's site on an active earthquake fault. Nevertheless, power generation began in 1986, and the plant has drawn very little public attention since.

Take Highway 101 north until Highway 1 branches off to the west in *San Luis Obispo* and follow Highway 1 north toward Morro Bay. Along the way you will pass by a series of conical

Morro Bay. Photograph by Bernard O. Bauer.

mountains that are distinctive in shape and color. A series of fourteen of these mountains—known as buttes or morros—extend from just east of San Luis Obispo to the offshore, submerged regions of Morro Bay, and in so doing, separate the Los Osos and Chorro valleys. Geologically, the morros are interesting because they are made of dacite (an igneous rock formed by volcanic activity), which has resisted the brunt of almost 25 million years of erosion, whereas the surrounding material (volcanic ash and soft sedimentary rock) has long since been transported to the sea. Thus, these looming figures are merely remnants of what were once a series of buried volcanic necks in which the molten magma stopped flowing and eventually solidified into what are called volcanic plugs. It is thought that these plugs were actually formed along an extensive fault line in the vicinity of Palm Springs, and over the eons and via the process of plate tectonics, they have since shifted to their present location. Nine of the peaks have been officially named. For example, Islay Hill just outside San Luis Obispo is named after the local wild cherry, Bishop Peak (the tallest, at 1,559 feet) has three peaks and resembles a bishop's headdress, whereas Hollister Peak was named after the family that ranched the land surrounding its base. Morro Rock (the shortest at 576 feet) was probably the first of the series to be sighted and named by Europeans because of its coastal location—Juan Rodriguez Cabrillo named it El Moro in 1542. Collectively, the nine peaks are known as The Nine Sisters.

As you approach the town of Morro Bay, turn left on South Bay Boulevard in front of Black Hill, the last of the landlocked volcanic plugs (watch for signs directing you toward Morro Bay State Park). Black Hill is within the boundary of the state park and is accessible to hikers. Turn right on State Park Road, which leads you through the campground (make reservations well in advance!) into a grove of eucalyptus trees, past a golf course on the right and the Morro Bay Museum of Natural History on the left. The latter is well worth visiting for a detailed account of the natural splendor of this unique location. Colorful displays and courteous attendants make this a treat for children and adults alike. There are some fascinating historical photographs of Morro Bay before development and of local economic activities, including 1930s photographs of mountains of abalone shells (unfortunately, commercial harvesting and sea-otter predation have drastically reduced the annual yield).

The eucalyptus trees are easily identified by their characteristic fragrance and stripped bark. The species was introduced to California in the 1800s in the hope that their rapid growth rate would supply the needs of an expanding railway system. The twisted and contorted nature of the dried lumber was not suitable for railway ties, but this was not recognized before the plant spread and established numerous rogue groves. Thus, eucalyptus can be found almost everywhere and they constitute some of the most scenic and aromatic forests along the coast of California. A particularly majestic grove grows south of Morro Bay on the route to Montana de Oro State Park. Not only can you see the eucalyptus forest, but also a splendid view of Morro Rock and the magnificent dunes along Morro Bay spit. A hike down Hazard Canyon offers some of the finest tidepool viewing along the coast.

State Park Road will eventually lead you into the road grid of the city of *Morro Bay*. When you reach Harbor Street, turn left, and it will take you toward the harbor and Embarcadero Street. You will immediately be struck by Morro Bay's most famous resident—the Rock. Since 1969, Morro Rock has been protected—the thirty-acre ecological reserve was established to prevent further destruction of the promontory from quarrying and more importantly, to serve as a nesting site for the endangered peregrine falcon. The Rock was once an island more than 1,000 feet from shore, and the entrance to the natural harbor was via a passage from the north. However, the resistant volcanic rock was a cheap and accessible source of construction material for the breakwaters of both Morro Bay and San Luis. More than a million tons of material were quarried before the reserve was established. In the interim, extensive efforts to clear the south channel for navigation, and the need for road access to the Rock produced the causeway (tombolo) that now connects the Rock to the mainland. The situation is not unlike that of Santa Barbara, where the extensive beach to the north of the artificial structure owes its existence to a blockage of the natural sand pathway along the coast (Leadbetter Beach in the case of Santa Barbara, and Estero Beach in the case of Morro Bay). Morro Rock is accessible by car via Embarcadero and Coleman Drive. The plentiful parking will allow you to enjoy some surfing, swimming, sunbathing, or a leisurely stroll out to the breakwater that protects the harbor. A walk on the breakwater is not recommended on stormy days when large waves may overtop the structure. In either case, a walk along the harbor side of the tombolo is usually rewarded with a sighting of a harbor seal or playful sea otter—they seem to enjoy the attention of their curious onlookers, but will often bark for a treat.

The harbor itself is home to two Coast Guard cutters and the only commercial fishing fleet between Santa Barbara and Monterey. There are 2,538 feet of piers, 6,934 feet of floating docks, and offshore moorings to accommodate 373 vessels of 45 tons or 65 feet in length. The value of the annual catch exceeds $2 million. In addition, there are many sport-fishing vessels and tour boats for the seaworthy. Of course, this is a place for fresh seafood, and a walk along the waterfront of Embarcadero will prove this true. Don't forget to try the saltwater taffy—a local delicacy for the sweet-toothed!

However, seafood and the Rock are not the only attractions in Morro Bay for the more than 1.5 million annual visitors. Thousands of bird-watchers flock to the area to spy on the more than 100 species of local and 250 species of migratory birds that are attracted to the wetland at the head of the estuary into Morro Bay. This 400-acre estuary is covered in eelgrass, and receives fresh water from Chorro and Toro creeks and salt water from the tidal currents that periodically inundate the winding channels. The sediments are home to many species of aquatic insects, invertebrates, and shellfish including cockles, clams, and oysters. Thus, there are plenty of culinary delights for birds to feed on. Egrets, kingfishers, pelicans, herons, terns, loons, teals, geese, ducks, buffleheads, wigeons, and, of course, peregrine falcons make their home either in the sheltered and misty estuary or on the dramatic cliffs that surround the region.

Birdwatching has become so popular that the entire town has been designated a sanctuary, and therefore no firearms of any kind are allowed within the city limits. In October, the skies fill with another flying creature—the monarch butterfly. These colorful insects appear in clouds of orange and gold as they make their way south for the winter. The eucalypt groves make a favorite stopping and feeding ground on their long journey. All in all, it is safe to say that Morro Bay has something for everyone.

△ Day Three

THE WILDEST COAST—MORRO BAY TO MONTEREY

The Tour

MORRO BAY TO CAMBRIA

Leaving Morro Bay, the route continues north on Highway 1. Visible to the west will be the sandy beaches and dunes of *Atascadero State Beach,* and later, *Morro Strand State Beach.* Both offer pleasant beach-combing and surf-fishing. Either would be good picnic spots, but there are many better spots within an hour's drive to the north. Several sets of buoys just offshore, usually in clusters of four, mark anchorage locations used by ships for transferring oil to storage tanks for the power plants.

To the east of Highway 1, numerous housing developments compete for remaining ocean views. The ongoing housing boom in this region is being driven by the expanding tourist industry and the demands of retirees seeking escape from the pressures of intense urban living.

About 5 miles north of Morro Bay lies the town of *Cayucos.* This is a former farming and fishing center, where tourism is becoming increasingly important for the local economy. Partly as a result, downtown Cayucos is experiencing a small-scale renaissance convinced by cafes and boutiques near the waterfront. Good views along the coast can be obtained from the recreation and fishing pier. A few miles north of Cayucos the highway turns inland toward rolling hills and grasslands, paralleling Perry Creek

Morro Bay to San Francisco

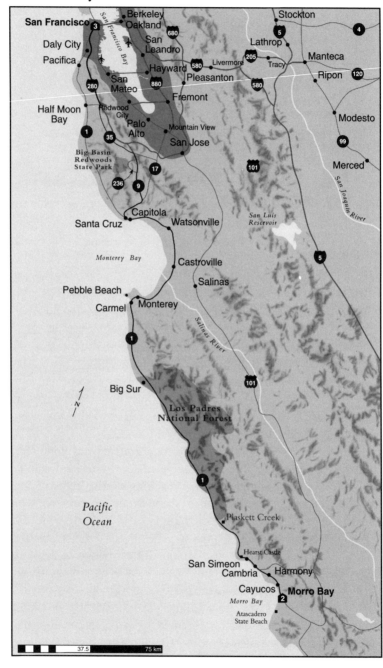

to the west. The effects of long-term grazing are apparent along this route: denuded slopes (in the summer) and the frequent scars of shallow landslides. The countryside here has escaped much of the development pressure common closer to the coast, and looks much as it did fifty years ago.

The small settlement of *Harmony,* a former utopian community, lies on the east side of Highway 1, about 9 miles north of Cayucos. After years of near-abandonment, Harmony has begun revitalization as an arts and crafts center. On the east side of Highway 1 the first stands of Monterey pine can be seen. The trees are relatively straight and tall here, but as we move farther north to more exposed locations, the Monterey pines become twisted and stunted.

Six miles from Harmony, is *Cambria,* the last substantial town until Carmel, about 100 miles to the north. Cambria has enjoyed a rich history, beginning with the establishment of a lumber mill in 1861. Originally known as Slabtown, acknowledging the rough quality of lumber used in local construction, the town's name was changed to Cambria in 1869. Cambria was the Roman name for Wales. In the middle of the nineteenth century, the town served as a local center for whaling and dairying. Hides, butter, and cheese were shipped almost daily for merchants in San Francisco.

In the nearby Santa Lucia Range, mercury and copper was mined. The rowdy life-styles of miners and lumbermen, once common in the streets of Cambria, have given way to the gentler habits of modern tourists. The central section of town, a short diversion off Highway 1, has many boutiques, galleries, and cafes. Crafts are the future in this delightful valley.

Follow Main Street through town and turn right on Cornwall Street, and then right again on Hillside Drive. Drive up the hill to consider the thought processes of "Captain Nitwit." The captain (Art Beal) was both architect and builder of the house on Nit Witt Ridge. This mélange of shell, wrecking yard refuse, wood, and stone is a California Historical Landmark. Perhaps Captain Nitwit fancied himself the poor man's Hearst.

Return to Main Street in Cambria, turn right, and proceed through the rest of the town. At the stop sign at the north end of Cambria,

THE SANTA LUCIA MOUNTAINS

The Santa Lucias are part of the Coast Range system that backs most of the central and northern California shoreline. The mountains are the result of folding and faulting along a northwest–southeast trend, roughly paralleling the San Andreas Fault in the *Santa Lucia Range*. This range is about 140 miles long, about 25 miles wide, with elevations approaching 5,904 feet. The mountains themselves are rugged and quite steep, plunging directly to the sea at many points along their western edge.

The basement rocks of the Santa Lucias are of two main types: the Franciscan assemblage and the Salinian block, the latter comprising the Sur metamorphic series and granitic plutons. The Franciscan assemblage is common throughout the Coast Ranges. As its name suggests, the Franciscan is a complex group of several rock types. Dominant in this group are metamorphosed sedimentary rocks, especially sandstones (graywackes) averaging 25,000 feet in thickness, interbedded with shale, limestone, and chert. The total unit is more than 50,000 feet thick. Units within this assemblage have been dated as early Cretaceous (about 120 million years old). Franciscan outcrops can be seen frequently along the roadcuts of Highway 1.

The graywackes are in gray-green beds up to 10 feet thick, separated by beds of darker shales. Limestones are apparent by their light coloration. Where the limestones are exposed near the ocean, the water is frequently tinted a distinctive light blue. Look for this near Limekiln Beach and at other lookout points. There are also exposures of Paleozoic (about 250 million years B.P.) limestones in roadcuts near Garrapata Creek.

The Salinian block is made up of crystalline rocks. The Sur series, portions of which date to the Paleozoic, is composed of gneiss, marble, quartzite, and schist. These can be seen at many locations in the Santa Lucias. The granitic plutons (smaller versions of those present in the Sierra Nevada) are late Cretaceous (about 90 million years BP), and made up of granodiorite and quartz diorite.

Because the Santa Lucia Range lies to the west of the San Andreas Fault, these mountains are part of the Pacific Plate. The northward movement of this plate relative to the North American Plate suggests that the basement rocks may have been deposited when the Santa Lucias (and other portions of the Coast Ranges) were as far as 350 miles south of their present location, where they may have been part of the Sierra Nevada Range.

The structure of the Santa Lucia Range is the result of faulting and folding. Folding, in particular, has been intense, complicating the interpretation of the geologic history of the mountains. Their present form is largely the result of the Pleistocene Coast Range orogeny (mountain building), about one million years BP. The extremely steep hillslopes of the Santa Lucias are characteristic of the relative youth of these mountains, as denudation processes have not had sufficient time to soften the appearance of the landscape. This is also why there are few large drainage basins in the range.

The steep hillsides and lack of river valleys severely restricted access to the area. At present, few roads cross the mountains between Cambria and Carmel, with only the Nacimiento–Fergusson Road traversing the central part of the range. Within the Santa Lucia Mountains are the 325,000-acre Los Padres National Forest and the more than 160,000-acre Ventana Wilderness Area.

KELP

Along much of the California coast, just outside the surf zone, we frequently see large patches of brown-red vegetation floating at or near the water's surface. This is the canopy of the kelp forest, part of an important coastal ecosystem.

The kelps are giant forms of brown algae (*Phaeophyta*). Strings of gas-filled bladders suspend the stipes ("stems") and leaves of the kelps in the water column. Waves and currents move through these forests like a silent wind, tugging, straining, and twisting these submarine trees. The canopy kelps that we see include giant, or bladder, kelp (*Macrocystis pyrifera*), bull kelp (*Nereocystis luetkeana*), and, nearer the shore, feather boa kelp (*Egregia menziesii*).

Other algae within this community include those that grow directly on the rocky substrate, and several other kelps that form layers of understory. Together, these algae provide the structure of a complex ecosystem. The rapid growth rates of the kelps (up to 10 inches per day) help the rapid cycling of nutrients. The fronds and stipes provide shelter to many species of invertebrates, fishes, and marine mammals. The holdfasts (root systems that bind the kelp to exposed rock) are home to crabs, worms, sea stars, and small fish.

Kelp is harvested commercially at several locations along the coast. The surface growth is "mowed" and collected by boat for processing. The kelp is used as a food, as a source of iodine and potash, and to produce algin (used in food-processing and the manufacture of cosmetics).

turn left, cross the highway, and turn right on Moonstone Beach Drive.

Moonstone Beach Drive leads to a small cove where a variety of diversions exist, including beach-combing, picnicking, and swimming. Coastal cliffs enhance the aesthetics of the local landscape. At Santa Rosa Creek, you enter the California Sea Otter Game Refuge. The otter's favorite habitat, kelp, becomes abundant in this section of the coast. Moonstone Beach was also the site of Leffingwell's pier, built in 1874 to serve Cambria in its commerce with San Francisco. Follow Moonstone Beach Drive along the coastal terrace until it rejoins Highway 1. Turn north. The elevation of the well-defined terrace through this area is between about 40 to 50 feet above sea level. The terrace remnants are now used mainly for grazing, although other agriculture was practiced in the past. These relict landforms provide just about the only near-level land along this coastline.

SAN SIMEON TO LUCIA

About 5 miles north of Cambria is the touristic village of *San Simeon*. The motels and restaurants are built across the small plain of the marine terrace. These facilities serve largely to enhance access to Hearst's San Simeon Castle. The original village of San Simeon is another 5 miles north. Old San Simeon was made famous as the small port William Randolph Hearst used to land materials and guests for "La Cuesta Encantada"—the enchanted hill. San Simeon Road will take you to both Hearst Castle, to the east, and into San Simeon proper, toward the west. Hearst Castle— truly a unique piece of architecture, landscape, and American culture—is well worth a visit. (All visitors must take the guided tour. Call ahead for hours, fees, and tour reservations: 1-800-444-7275.)

About 3 miles north of San Simeon, the *Piedras Blancas Lighthouse* can be seen in the distance to the west of Highway 1. The structure is more than 100 years old (1874) and is a major navigational aid along this lonely, frequently foggy coast. *Piedras blancas* means white rocks in Spanish; seabirds have stained the small rocky islands white with guano. Small cliffs back the beach along

William Randolph Hearst's other "castle": the Hacienda/Milpitas House, designed by Julia Morgan, at Fort Hunter-Liggett, north of San Simeon. Photograph by Paul F. Starrs.

this reach, and in places, strong winds have blown sand up and over the cliffs to form small dune fields.

Seven miles farther, beyond the turn for Piedras Blancas, is *Ragged Point,* where there is food, gas, and lodging. This is a good spot for beach-combing, with relatively easy beach access, small sandy beaches, cliffs, and rocky outcrops. About 1 mile north of Ragged Point is San Carpoforo Creek, where Gaspar de Portola encamped with his small band of soldiers during their overland expedition from San Diego to San Francisco, in the summer of 1769. Their natural highway, the marine terrace surface, ends here. On their return trip that winter, the Spanish soldiers again crossed the Santa Lucia Mountains. They survived by eating their mules, and supplemented that diet with fish provided by the Indians. Even the local natives avoided large sections of the steeply cliffed coast.

Sea arches carved by waves in rocks off Big Sur. Photograph by Bernard O. Bauer.

This route north, from San Simeon to Carmel, was not completed until 1937, eighteen years after it was begun. This was partly a "make-work" project during the Depression, although the original construction began earlier. Much of the labor was provided by convicts working their way toward reduced sentences. The road will not likely be widened substantially. Even now, the California Department of Transportation (CalTrans) works hard to keep the present lanes open, especially during heavy rains. After the extreme rainfalls during the winter of 1982 to 1983, landslides closed the road to through traffic for more than a year. This is one stretch of the journey where haste is futile. Enjoy the scenery—it is among the most spectacular in the world. Note that state parks are frequently designated for day use along this route, many providing good access to beaches and picnic facilities. Several small settlements sell food and gasoline. Campsites are also available, but space is at a premium throughout the summer. It would not be inappropriate to allow at least a half a day for the 100-mile drive—longer if possible!

Just north of Ragged Point we enter Monterey County. The road here closely hugs the slopes above the ocean. Landslides and rock falls are common. Be especially cautious at night, in inclement weather, and early in the morning. Local instabilities and previous landslides are evidenced by the downward tilting fences at the side of the road and the frequently repaired stretches of road.

The mountains inland are largely held by the federal government. The two largest units are the Los Padres National Forest and Hunter-Liggett Military Reservation. The latter was purchased from the vast Hearst properties during World War II.

On this stretch you will pass *Redwood Gulch,* about 7 miles north of Ragged Point. This marks the southernmost extent of the coast redwood (*Sequoia sempervirens*). Specimens of this tree may be seen within the protection of the incised coastal valleys north of this point. *Gorda* is about 2 miles north of Redwood Gulch.

Four miles north of Gorda is *Jade Cove,* named for the green serpentine found there. You can park at the Willow Creek vista point. There is also a smaller parking area at the base of the cliffs. Note that collecting is permissible, but only below the high-water line. Camping is permitted at Plaskett Creek, about 0.5 mile north, and there is a day-use facility at Sand Dollar Beach just a bit farther north.

Limekiln Creek, 5 miles north of Plaskett Creek, offers sightseeing, camping (the private campground, Limekiln Beach Redwoods Campground), and hiking; there was no day use of the campground in August 1991. A steep hike up Limekiln Creek rewards the hardy with more than ten plant communities, including alpine forest. The average grade up this valley exceeds 25 percent. At the mouth of Limekiln Creek is the abandoned Rockland Landing, and a short distance away, ruins of the kilns. Lime, used in construction, was shipped from this poor harbor in the 1870s.

LUCIA TO CARMEL

The Lucia Lodge, another 0.5 mile along the highway, offers lodging, a good restaurant, gas, and groceries. Thirteen miles north you will pass the entrance to the Esalen Institute. Built around a series of hot springs, the isolated institute has long been a hub for workshops exploring the psycho-potential of personality. It is open to the public, but only in the wee hours of the morning, literally (1:00 to 5:00 a.m.). You can cook your persona, and then continue north.

The sign-posted vista point, about 15 miles north of Lucia, is a superb photo location. Waters to the south are frequently tinted with the outwash from limestone deposits. A series of plaques describe the ecology of the area, including monarch butterflies and whales.

The Coast Gallery and the Henry Miller Memorial Library are about 21 miles and 24.5 miles north, respectively, of Limekiln Creek. Local artists and artisans display and create at the Coast Gallery, and there is an exhibit of some of Henry Miller's work. Miller lived in the Big Sur more than fifteen years, and the Memorial Library recalls part of his experience. Letters, photographs, and books are on display. Between these cultural outposts begins a thicket of restaurants, lodges, and inns. The quality of hospitality varies, but many of the establishments, and views, are first-rate.

Near *Nepenthe* (named after a mythical Greek drug that induces the forgetfulness of sorrows), the highway turns inland to the Big Sur Valley. The Big Sur River is the largest stream on the western slope of the Santa Lucia Mountains, and its valley has been one of the principal access routes to the interior of the range. The valley was home to much of the earliest settlement in the region, and long occupied by Esselan Indians—note name similarity with the Esalen Institute—one of the first of the California tribes to become

MARINE MAMMALS OF THE CALIFORNIA COAST

In the middle of the nineteenth century, San Francisco was the hub of a thriving whaling industry. More than 500 vessels swept the coastal waters between Alaska and Baja California to meet the demand for whale oil. By 1881, however, the industry had virtually collapsed, and only about forty ships remained. The intervening years were times of prodigious slaughter.

Captain Charles Scammon, based out of San Francisco, amazed his competitors with the speed at which he obtained his ship's capacity of oil. Whalers following his ship soon realized the source of his bounty: Captain Scammon had discovered the winter breeding grounds of the *gray whale*. Falling upon the whales in the shallow waters of Scammon's Lagoon, in Baja California, the hunters took numbers that far exceeded the natural replenishment rate. By 1890 the gray whale was believed to be extinct.

Today, these magnificent creatures once more make their 10,000-mile annual migration along the Pacific coast. They can be observed from rocky headlands from Point Reyes to Point Loma, and a fleet of small craft, from ports all along the California coast, now chase the gray whales to bring them within camera range of the curious. Although the whale remains on the endangered species list, its comeback is one of the success stories of marine-mammal management via international treaty.

A similar story accompanies the near-extinction of the sea otter in California. The Russian settlement at Fort Ross (in present-day northern Sonoma County) was the base of an otter "fishery" that quickly destroyed the population north of San Francisco Bay. Attention then turned to the coastal waters as far south as San Pedro.

Sign on Seventeen-Mile Drive shows the value of sea mammals to the elite property-owners on the coast. Photograph by Martin S. Kenzer.

The focus of this activity was the luxuriant pelt of the southern sea otter. In the early 1800s this fur fetched between fifty and a hundred dollars in the markets of China. As many as 18,000 pelts were delivered to Canton in a season. The hunt occurred within sheltered harbors, along rocky coasts, and within the kelp beds where the otters sought food. Demand for these furs also drove the *sea otter* to the edge of extinction in the 1800s.

Currently, the waters along the California coast are home to more than thirty species of marine mammals, at least nine of which are on the federal endangered or threatened species list. All of these animals are protected under the Marine Mammals Act, and several species are further protected through their status as threatened or endangered species.

Cetaceans are well-represented in these waters. Examples include the humpback whale (*Megaptera novaeangliae*) off the shores of southern California; the gray whale (*Eschrichtius robustus*) that migrates from one end of the state to the other in its annual pilgrimage; the largest living animals, the blue whale (*Balaenoptera musculus*); and the sperm whale (*Physeter macrocephalus*), the source of ambergris, a perfume base.

Smaller whale species include dolphins and porpoises that are common in many of the coastal waters. These include the bottle-nosed dolphin (*Tursiops truncatus*) and the Pacific white-sided dolphin (*Lagenorhynchus obliquidens*), both frequenting the southern and central waters of the state.

The *California Sea Otter Game Refuge,* essentially coincidental with the Monterey County coastline, is home to a burgeoning population of the southern sea otter (*Enhydra lutris*), and they can be frequently observed in kelp beds, especially along the Big Sur coast and at Point Lobos, south of Carmel. These otters are tool users: They employ small rocks to crack open shellfish and crabs. They have voracious appetites, daily consuming about one fourth of their body weight. Sea otters may also be seen sporting in the waters of Morro Bay near the inner breakwater.

Also common along the coast are members of the order Pinnipedia: seals and sea lions. *Harbor seals* (*Phoca vitulina*) are likely to be seen in several of the sheltered waters. In particular, look in Morro Bay, Point Lobos, at several of the lookout points along 17-mile drive between Carmel and Monterey, and Monterey Bay near Cannery Row.

Less common, but perhaps more dramatic, is the *elephant seal* (*Mirounga angustirostris*), which may weigh more than 2 tons and be more than 15 feet long. Elephant seals breed along the central coast in the winter. Also present along the central coast is the *Stellar sea lion* (*Eumetopius jubatus*) that exceeds lengths of 12 feet and weights of about one ton. More common along most of the shore south of San Francisco is the *California sea lion* (*Zalophus californianus*). These are the performing "seals" trained for amusements. Their barking can be heard most mornings from quieter coastal waters such as those in the shelter of the Monterey Peninsula.

extinct. With the present tourist overloading, it is difficult to recapture the sense of isolation and grandeur once presented by the deep shadows beneath the redwoods and the near-silence of daybreak.

With the advent of farming and mining in the middle of the nineteenth century, there was a rapid expansion of the Hispanic and Anglo populations in this valley. There was also an active lumber industry carried out mainly near the coast. Monterey jack cheese was invented in this valley in the 1890s, as a dairy product aimed at the markets in Monterey and San Francisco.

This section of the *Big Sur* region has also long been a focus for counter-culture artists and authors, as exemplified by Henry Miller, Jack Kerouac (somewhat of an imposter here), Richard Brautigan, and Robinson Jeffers. The area was also a more natural alternative to Haight-Ashbury and the Sunset Strip as a magnet for the youth of the 1970s.

As the road drops nearer the valley floor, in the vicinity of the River Inn, thick stands of oak and sycamore occur. These trees thrive in the fertile soils next to the river, where abundant water is also available. As the valley opens toward the sea, these trees closely hug the Big Sur River, on the left. The entrance to Andrew Molera State Park is here also. Molera was one of the pioneering agriculturalists in the valley.

About 4 miles north of Big Sur, the Big Sur lighthouse will be visible in the west. The lighthouse and outbuildings occupy a rocky promontory that was once an island. Sands, transported by waves and currents, have been deposited in the shelter of the island, creating a tombolo similar to that at Morro Bay. The wind has blown additional sand onshore to form dunes.

A few miles farther, Highway 1 runs between sets of sand dunes that have been blown up and over the cliff from the beach below. The source of this sediment is the beach formed at the mouth of the Little Sur River, about 5 miles from the Big Sur lighthouse. Coastal scrub vegetation is especially luxuriant along this stretch of highway. Species include coyote brush, black sage, laurel sumac, ceanothus, and lupine, for example.

Four miles farther is *Bixby Creek* and the rainbow bridge that spans it. Completed in 1932, the bridge was one of the major links in completing the highway through the region. The Rainbow Lodge was once immediately north of the bridge, an important road house and traveler's rest. Turnouts here afford good sightseeing. Note, especially, the limestone outcrops at about the level of the beach on the south side of Bixby Creek. There is also a small, but nicely formed, sea cave in the cliffs just seaward of the overlook.

About 7 miles farther, we begin seeing the first fringes of development around Carmel. Houses here tend to be very expensive and to occupy large lots. Watch on your left (west side of the road) for the entrance to Point Lobos State park, a jewel of natural beauty and historic interest. Each stretch of coastline in this small park illustrates a different twist of coastal geology and biology. Not far beyond Point Lobos (named for the sea lions—"wolves"—who live here), Highway 1 sweeps down to the ocean at a small embayment known as Monastery Beach. This is a popular scuba-diving site because the beach is quite steep, aiding movement between the water and the beach. The beach is named after the monastery across the highway.

Just to the north is the Carmel River Valley. The intensive agriculture in this valley is somewhat anachronistic. The sight of artichoke harvesters working against a backdrop of some of the region's most expensive real estate exemplifies the often schizophrenic nature of the developing California landscape. The soils in the Carmel River Valley are quite fertile, supporting high dollar-yield crops, such as artichokes. The Carmel River flows through this portion of its course only rarely, although it did make it to the ocean in 1991, as a result of intense March rains.

Turn left at Rio Road (at the first traffic signal), and follow the avenue into *Carmel*. You will pass the Mission San Carlos Borremeo (Carmel). The mission was established in 1770 by Father Junipero Serra. The mission is still active and the grounds beautifully kept; it is well worth a visit.

To tour Carmel quickly, continue along Rio Road to Ocean Avenue and turn left. This is the heart of central Carmel. If you are in the mood for shopping, dining, or strolling, find a parking place and enjoy the village atmosphere of this delightful neighborhood. If you continue down Ocean, you will come to Carmel City Beach. The brilliant white sands and Monterey cypress against the sky and sea make this a terrific picnic and photo spot.

CARMEL TO MONTEREY

Leaving Carmel Beach, drive up Ocean Avenue to North San Antonio Avenue, and turn left. This road becomes Carmel Way, and leads to scenic *17-Mile Drive*. Seventeen-mile Drive is privately owned—recently purchased by a Japanese entrepreneur—and there is an admission fee. Some of the most prestigious golf courses in the world, including Pebble Beach and Spyglass Hill, are here. Stately mansions hide in the trees lining the streets. The visitor will be given a guide to points of interest, including the Ghost Tree, the Lone Cypress (one of the most photographed trees in the world), Cypress Point Lookout, the malodorous Seal and Bird rocks, and Point Joe, the false entrance to Monterey Bay.

Exit 17-Mile Drive at the Pacific Grove Gate, and turn left on Sunset Drive, or continue the drive to the interior of the Del Monte Forest. We follow the former route. Sunset Drive follows the coast around the contour of the *Monterey Peninsula*. Enjoy views of blue water, white surf, plentiful bird life and, perhaps, otters and seals.

At the northern end of the peninsula, you will enter *Pacific Grove*. Established as a religious colony, and used as a foil for the irreverent Steinbeck novels set along Cannery Row, the town presents a well-preserved facade of Victorian architecture, boutiques, and bed-and-breakfast establishments. It is also the upscale guardian to the western edges of Cannery Row and the Presidio.

The Lone Cypress, Seventeen-Mile Drive, Carmel. Photograph by Barbara Lerner Kopel.

IN AND AROUND MONTEREY BAY

The Monterey Bay area is one of the most picturesque expanses along the California coast. Dotted with Hispanic towns first settled in the eighteenth century, and mingled with Anglo frontier towns founded a century later, the area in and around Monterey Bay is delightful, charming, and above all a captivating place to live. The population of the region has roughly doubled during the past fifteen years (to 1.5 million), fueled by invasions of commuters, retirees, and those seeking alternative lifestyles. Once a quiet, slow-growth area, the entire Monterey Bay region finds itself today tied by concrete umbilical cords to the San Francisco Bay region, and is under pressure to develop suburban lifestyles.

Along with suburban growth, however, is equal pressure to alter what is authentic, to cash in on the increasing demands of tourists. The city of Monterey, for instance, has become a prime tourist draw for northern California. Once the capital of Spanish California, and long-characterized by low-lying Spanish colonial architecture, by the 1960s Monterey had become a small fishing town with a quaint waterfront, a seedy but historic cannery row, and a sweeping view of Monterey Bay. Fishing declined and gave way to resort hotels and conventions, and Spanish architecture was replaced by "institutional" construction. Today, Monterey is a confusing city, with elements of yesterday hidden among the gaudy tourist trade of the 1990s.

Carmel is Monterey's sister city, nestled among the pines along Carmel Bay. Unlike Monterey, however, Carmel is first and foremost a tourist city; it is also one of California's most unusual cities. Carmel has approximately 6,000 permanent residents, but on a busy day, more than 8,000 cars will be parked on its streets. It is a wealthy city, which was

Cannery Row, Monterey waterfront. Photograph by Paul F. Starrs.

established as an artists' colony around the turn of the century. For those who live in Carmel, the tourists are both a blessing and a nuisance. The statue of Junipero Serra—the Franciscan priest who founded the area and is generally considered the Father of the California missions—for many years had the words "tourist go home" painted across it. Today, Carmel has become the home of the nouveau riche, mostly from San Francisco. It remains an artists' colony, but only the most successful artist can afford to live within this upscale community. In some ways, it is a Disneyland for the rich tourist, crammed full of shops that sell expensive clothing, antiques, and art.

Less famous towns along Monterey Bay are *Salinas* and *Watsonville,* which are almost always passed over by tourists because they are chiefly agricultural centers that offer few tourist amenities. John Steinbeck, who chronicled the arduous life of migrant farmworkers in the 1930s, was born in Salinas; his childhood home and the Steinbeck Library are both prominent buildings in downtown Salinas. The majority of Watsonville's and Salinas's present-day populations remain Hispanic, migrant laborers. Both towns are known across the state for their agricultural crops, most notably artichokes, apples, lettuce, brussels sprouts, celery, and cauliflower, all of which are sold throughout North America. Cutting into the fields, however, are new subdivisions for people willing to trade a two-hour commute to Bay Area jobs for affordable housing.

The city of *Santa Cruz*—both a mission and a pueblo during early Spanish occupation of California—languished throughout the nineteenth century, during which its only economic base was lumbering of the giant redwoods in the Santa Cruz Mountains. With the turn of the century, a small railroad spur was pushed into Santa Cruz, thereby linking it with the greater San Francisco Bay area. Once the railroad was in place, Santa Cruz became a favorite beach resort and summer playground for San Franciscans; the area was coveted for its beautiful, sandy beaches, the mountain views of Monterey Bay, and the high surf that crashes directly onto a nearly vertical shoreline. Until suburban pressures of the 1980s, Santa Cruz remained a small, quiet coastal community, known largely for the University of California campus tucked away in the nearby redwoods, its many Victorian homes, and its penchant for embracing alternative lifestyles. Today, however, Santa Cruz is a bedroom community for those who work in the San Francisco Bay area. In 1989, it was hit by a sharp earthquake, which had devastating effects on its turn-of-the-century downtown.

MONTEREY TO SALINAS

Leave Monterey via State Highway 68 east, toward Salinas. Route 68 can be picked up off Highway 1 near the Naval Postgraduate School at the east end of town. Highway 68 passes between the Old Del Monte and U.S. Navy golf courses, and just south of the Monterey Peninsula Airport.

This road leads through prime California oak grassland, used primarily for grazing here. When the fog is in, it is easy to imagine the old California of rancho days. After about 7 miles, Laguna Seca (dry lagoon) Recreation Area, including the world-renowned race track, appears on the left. Immediately thereafter, the left side of the road bounds Fort Ord Military Reservation. Long one of the principal West Coast military establishments, Fort Ord is currently scheduled for closure under Defense Department cost-cutting.

About 5 miles beyond *Laguna Seca,* Highway 68 drops off the mesa, much of which is composed of ancient sand hills, down the edge of Toro Creek. Directly ahead is the alluvial plain of the Salinas River. This is some of the richest agriculture land in the state, and evidence of intense farming is visible everywhere.

At the base of the bluff, Highway 68 crosses River Road. One mile farther, turn right onto Spreckels Boulevard.

This long, tree-lined boulevard leads into the old company town of *Spreckels.* Settlement here was driven by the presence of the sugar-beet fields and associated processing facilities. The Steinbeck Library and the author's boyhood home are both within walking distance of the downtown. Return to Highway 68 and go on through Salinas.

SALINAS TO SANTA CRUZ

Leaving Salinas, follow Highway 183, the Watsonville Highway, northwest toward *Castroville,* the self-proclaimed Artichoke Capital of the World. And, indeed, evidence supporting this claim is abundant in the fertile fields along this highway. Artichokes are members of the thistle family, and the association is visible in the spine-tipped leaves. (It is hard to imagine the process whereby the edibility of artichokes was established.)

SUGAR AND THE MONTEREY BAY REGION

During the latter part of the nineteenth century, farmers throughout the Monterey Bay area were struggling to find a commercial crop that would take advantage of local climate and soil conditions, as well as command a premium price in the marketplace. Potatoes, a variety of green vegetables, and a host of field crops had been tried, but without a great deal of success. High transportation costs and competition from farmers closer to markets proved to be significant barriers.

Then, in 1888, Claus Spreckels proposed to build a beet-sugar factory in Watsonville and began to refine sugar for sale. Most sugar in the United States was imported from Europe at the time, where German scientists had developed methods to refine the beet to produce sugar profitably. By 1889, more than 3,000 acres of sugar beets were planted in and around the Watsonville area. To accommodate the new crop, rail lines were constructed to connect the Monterey Bay area with the port of Oakland; from Oakland, most of the sugar was sent to the expanding American Middle West.

By 1890, 1,000 tons of beet sugar were produced by the Western Sugar Beet Company, and by 1892 the rich alluvial soil throughout the region had helped yield the largest crop of beets in the United States—beets containing a higher percentage of sugar and of a superior quantity than those of Germany. The factory processed an average of 350 tons of beets each day, resulting in 45 tons of usable, refined sugar. By 1895, 11,000 acres of beets were planted, and in 1897 155,000 tons of beets were cut, producing 20,000 tons of refined sugar.

The factory in Watsonville was sold in 1898 to the American Sugar Refining Company; construction on a new factory, in present-day Spreckels, began immediately. The railroad was extended to Salinas and later to Spreckels, to serve the new plant. The city of *Spreckels* was built as a company town for employees working at the new refinery. The refinery remains today as an important producer of sugar, and Spreckels continues in the tradition of a company town, reflecting its late-nineteenth-century origin.

Castroville is an agriculture-based community, occupying a small rise above the surrounding plain. Just past the intersection of Highway 183 with Highway 156, the Giant Artichoke restaurant is on the right. The menu will quickly convince you of the culinary versatility of the artichoke. They even offer artichoke wine in the gift shop.

Other unusual eateries in town include Scalia's Italian, Bing's Diner (great for breakfast or ribs), and the Central Texas Bar-be-que. Ask one of the locals to tell you about Marilyn Monroe's reign as the Artichoke Queen.

Follow the main street through Castroville, and merge with Highway 1 north at the edge of town. Visible to the northwest are the high stacks of the Pacific Gas and Electric generating station at Moss Landing. At the southern edge of town, take Moss Landing Road, forking toward the left. This is the old highway. It leads to one of the few, non-beautified seaports along the central California coast. This is no cute fisherman's wharf; this is where people pry a living from the sea's resources. Boat yards, fish processing, and an earthy citizenry dominate the scenery. Follow the road past a series of antique stores, some of which still offer undiscovered treasure. Rejoin Highway 1 in the shadow of the Whole Enchilada restaurant and bar. Live music and good seafood are attractions here.

Follow Highway 1 north, across Elkhorn Slough, a former mouth of the Salinas River. Pleasure craft of all sizes and shapes are more common in this end of the harbor. The road now leads over a series of rolling hills, still intensively cultivated; brussels sprouts are common. The farmlands of the lower Pajaro Valley include large undeveloped tracts that are part of the Pajaro Valley Wetlands. These freshwater marshes and waterways represent a substantial natural resource for their floral diversity and avian habitat. Enjoy the drive toward Santa Cruz, about 30 miles to the north.

Several of the coastal communities south afford excellent beach recreation. To avoid the crowds in Santa Cruz, try La Selva Beach or Capitola, both sign-posted off Highway 1. *Capitola* is one of the oldest resort communities in this region, and several generations of distinctive architecture survive, including Victorian and art deco structures. Most of the smaller villages in this area offer bistro-type dining as well as off-beat gift shopping.

SANTA CRUZ TO SAN FRANCISCO

Highway 1 runs straight through Santa Cruz. The tour will turn inland onto Highway 9, but first explore this unusual college and commuter town. Three turnoffs from Highway 1 are well worth taking. The Ocean Street turnoff goes down to the beach, board-walk, pier, and amusement park. Watch for the surfing museum! At the northern edge of town, look for the exit for Natural Bridges State Park, where you can walk along a trail overlooking spectac-ular holes and bridges carved through cliffs by ocean waves. Finally, a well-marked turnoff about 2 miles north of Santa Cruz takes you to the University of California, Santa Cruz, the most unconventional campus of California's prestigious university sys-tem, in a setting amid the redwoods that a resort would envy.

There are three possible routes north out of Santa Cruz. Travelers in a hurry should take the fast, unscenic route out of Santa Cruz (that is, leave Highway 1 for Highway 17 to San Jose, and then take either Interstate 280 or Highway 101 north to San Francisco). The spectac-ular coastline scenery, with possible glimpses of migrating whales, continues if you decide to follow Highway 1 northward. We describe the third route, which goes inland up the spine of the Peninsula—a very beautiful and leisurely trip, starting on Highway 9.

Leave Santa Cruz on Highway 9 north. The road follows the San Lorenzo River Valley back into the mountainous interior of Santa Cruz County. In the deeply shaded valley, a cooler, damper microclimate prevails. Redwoods closely crowd the valley edges. Trees are dusted with mosses and lichens. The forest floor supports

BIG BASIN REDWOODS STATE PARK

At Boulder Creek you can take a side-trip to Big Basin Redwoods State Park, via Highway 236. It is about 9 miles of sharp curves to Big Basin, and the redwoods are worth the effort. Look carefully in shady patches, you might see a banana slug, an unlikely creature attaining lengths approaching one foot (and mascot of the University of California, Santa Cruz). For the hardy driver, follow 236 north to rejoin Highway 9 after a serpentine journey. If you are driving a car with a trailer or recreational vehicle, return to Highway 9 via Boulder Creek. Highways 9 and 236 rejoin near Waterman Gap. It is another 6 miles to the intersection of Highway 9 with Highway 35. Turn north (left) onto Highway 35, Skyline Boulevard.

ferns and rustic cabins. Five miles from the Pacific Ocean you seem to be in the embrace of remote mountains.

The Santa Cruz Mountains have also escaped large-scale commercial or residential development because of their steep slopes and the limited access resulting from the dearth of major rivers. Instead, the quiet hills afford refuge to the wealthy and the poor: people preferring privacy to easy access.

The Santa Cruz Mountains were one of the earliest forestry regions, where extensive stands of coast redwoods were felled to supply the materials to build California's cities. This area also contains some of the earliest resort development in the region.

Highway 9 north through the mountains is curvy and full of wonderful views to be savored. In the first 13 miles along Highway 9, you will pass through the beautiful Henry Cowell Redwoods State Park, and then through the resort villages of Felton,

FILOLI GARDEN ESTATES

Consider a diversion to Filoli Garden Estates, in the community of Woodside, at the southern end of the Crystal Springs Reservoir. Take Highway 84 east, off Highway 35, to reach Woodside. The name Filoli is a composite from the words FIght, LOve, and LIve, taken from the credo of the original owner, William Bourne II. The house was completed in 1919, and is currently on the National Register of Historic Places. The house has 36,000 square feet of living area, in more than forty rooms. Fans of the television series "Dynasty" will recognize the elegant brick exterior of this mansion. Plan on spending a couple of hours to tour the estate.

Ben Lomond, and Boulder Creek. To stay or eat at the charming inns, lodges, and restaurants that line the road along the San Lorenzo River, reservations are strongly recommended.

At Saratoga Gap, you will leave Highway 9 for Skyline Boulevard (Highway 35) north. Skyline Boulevard follows the ridge-line of the Santa Cruz Mountains and lets you look eastward down into the crowded cities of the southern half of the San Francisco Bay.

The road is frequently mist-swept, and vistas may be obscured by low clouds or fog. However, when the skies clear, dramatic panoramic views are available from the numerous turnouts along the route. From the southerly view points, one may look east across the southernmost wetlands of San Francisco Bay to the mountains of Santa Clara County, including Mount Hamilton (4,209 feet), site of the *Lick Observatory.*

Look for the intensely developed central business districts of the cities of, from south to north, San Jose, Santa Clara, Sunnyvale, Mountain View, and Palo Alto. The runways and large hangars at Moffett Field Naval Air Stations, near Mountain View, are especially noticeable landmarks.

After about 15 miles, views will include the Dumbarton Bridge, linking Redwood City with Fremont, across the bay, and the much longer San Mateo Bridge that crosses the bay to Hayward. The developed downtown of *San Mateo* is near the proxial end of the bridge. Toward the northeast, in the distance, Mount Diablo (3,849 feet) may be visible. Mount Diablo is a major national survey base because it is prominent relative to the Bay Area and the central valley.

An excellent view area appears near the intersection of Skyline Boulevard with Highway 92. This point is just opposite San Mateo. Below the lookout, to the east, runs the major trace of the San Andreas Fault. Highway 280 is visible as it follows the rift zone north toward San Francisco. Crystal Springs Reservoir fills a portion of the valley created by movement along this fault. At this point, you are not, tectonically, on North America, but rather on the edge of the Pacific Plate.

Otherwise, turn left (west) on Highway 92, from Skyline Boulevard, for the 5-mile drive into Half Moon Bay. The road follows the valley of Pilarcitos (little pillars) Creek. Obester Winery awaits those tempted by award-winning wine. The wide coastal terraces support grazing and high-return crops such as tomatoes, brussels sprouts, and pumpkins, and had the first commercial artichoke plantings in the state.

The city of *Half Moon Bay* is centered just south of the intersection where you leave Highway 92 and turn north on Highway 1. It was originally known as San Benito, and was early developed as an agricultural center, especially for grazing. The settlement served as a support center for fishing and whaling. It is now experiencing additional development pressures as the far southwestern edge of the San Francisco commuter belt, and as a tourism and recreational site. The Great Pumpkin and Art Festival, held here each October, gives prizes for the largest pumpkins. In 1991, the winning specimen weighed more than 600 pounds.

Pillar Point Harbor is about 3 miles north of the city, just west of Highway 1. During the latter half of the twentieth century, whalers of Portuguese descent based here for the hunt for gray and humpback whales. There is a small "fisherman's wharf"-type development and both sport- and commercial-fishing fleets. In addition to restaurants, there are several good picnic spots along Highway 1 to the north of Pillar Point, including near Moss Beach, Montara Point, or Montara Beach.

About 2.5 miles north of Pillar Point Harbor, near Moss Beach, you will see signs indicating the *James V. Fitzgerald Marine Reserve*. A stroll along the beach at this reserve is a rewarding experience for those interested in the diversity of tidepool ecosystems. Alternating stretches of sandy and rocky shoreline offer an aesthetically challenging landscape.

About 1 mile farther along Highway 1, note the small lighthouse at Montara Point. This diminutive structure was built in 1928, and is now a National Historic Place. Adjacent is the Montara Lighthouse Hostel.

Leaving Montara, and driving north on Highway 1, additional beach access is available at *Montara State Beach* and *Gray Whale Cove State Beach*. On the latter beach clothing is optional, and access requires some stair climbing. To the north of this beach, and on the west side of Highway 1, a former World War II observation station for coastal defense is visible atop a steep ridge crest. The high degree of exposure of this structure, including much of its foundation, reflects the rapid rates of slope degradation that occur along this section of the coast.

Brace yourself as Highway 1 climbs across the cliffs of *Devil's Slide*. Here, steeply dipping beds of Paleocene shales and sandstones provide tenuous support for the roadbed. Earthquakes and rain are both hazards to this section of highway, and temporary closures, especially in winter, are common. After the traverse of Devil's Slide, the road drops back down toward sea level and the city of Pacifica.

This marks the return to a much more intensely settled landscape, another part of the San Francisco commuter-belt. *Pacifica* is a community founded in the late 1950s through the amalgamation of nine small communities. Continue north on Highway 1 through the suburban community of Daly City, built directly astride the San Andreas Fault. San Bruno Mountain is visible to the east, marking the southern boundary of the city and county of San Francisco. Highway 1 will become Nineteenth Avenue, and leads north to the Presidio and the Golden Gate Bridge, and the driving tour of the city.

△ Day Four

SAN FRANCISCO

San Francisco presents a seeming labyrinth of contrasting geographic images. San Francisco is cosmopolitan and, for some, feels like a European capital, yet it is the most Asian of our cities, with more than 200,000 of its residents (27 percent) claiming a trans-Pacific heritage. For emigrants from "Back East," San Francisco was, and still is, a promised land at the end of the continent. But, to generations of Asians, the city lies at the far eastern edge of Asia, serving as a gateway to a new life in North America.

Tourists also experience two apparently conflicting cities. Tourism is the primary industry, and city officials do everything they can to help you enjoy your visit. But the city's notorious hills, and the relentless road grid draped over them—not to mention the one-way streets—make touring a challenge.

Another example, the weather, is best described paradoxically as extremely temperate. It never freezes and summer fogs typically do not admit a heat wave. Yet, Mark Twain's oft-quoted complaint (probably apocryphal) that the coldest winter he ever spent was a summer in San Francisco serves as fair warning to visitors: bring warm clothes, whatever the season!

Finally, San Francisco has a reputation based on certain "signature" social characteristics—ethnic diversity, nonconformity, liberal politics, artistic creativity, and care for the poor. But, many U.S. central cities have developed ethnic and nonconformist districts, so in this and other respects San Francisco appears not so unique. Also, the black and Hispanic poor have been removed from much of the city by the economic forces of white gentrification, leaving San Francisco with average housing prices second only to those in New York's Manhattan.

Three lines of hills cross the city. The northernmost, running due east from the hills above the Presidio, includes the most famous ones: Pacific Heights, Russian Hill and Nob Hill, and finally Telegraph Hill with its distinctive Coit Tower. Another line can be visualized as an extension of outer (the southwest end) Market Street through the Castro District. The higher, but less famous outcrops at Corona Heights, Twin Peaks, and Mount Davidson extend southwesterly from Market. Finally, another line close to San Francisco Bay and U.S. Highway 101 consists of Bernal Heights and Potrero Hill. All nine peaks can be seen from the top of Corona Heights.

San Francisco is characterized by change. You will notice dramatic differences among and within neighborhoods as San Franciscans continue to remake their city in their own image. "The City" presents an excitement and vitality that is unique in North America.

The Tour

THE PRESIDIO

The tour starts at the Ayala Vista Point at the north end of the *Golden Gate Bridge,* where there is a good view of the city and its geographic context. Observe the city from afar and note the juxtaposition of Alcatraz Island before crossing the bridge again back into San Francisco. Take Highway 101 South and follow Presidio signs to the exit onto Lyon Street, heading south. Turn right on Lombard Street to enter the Presidio at the main gate (east entrance). After you explore the Presidio, leave again by the same gate.

The Presidio marks the beginning of San Francisco. As colonists on the eastern seaboard were declaring independence from British rule in 1776, the Spanish moved into San Francisco and set up Mission Dolores and a military command post here. The

San Francisco Tour

Golden Gate Bridge

To Sausalito

101

1

The Presidio

Marina

Ghirardelli Square

Fisherman's Wharf

Russian Hill

Nob Hill

North Beach

Telegraph Hill

Coit Tower

The Embarcadero

Transamerica Building

Financial District

Chinatown

Union Square

Van Ness Av

Market St

South of Market

Mission

Castro

Potrero Hill

280

101

Alcatraz Island

Treasure Island

San Francisco-Oakland Bay Bridge

80

San Francisco Bay

Union St

Pacific Heights

California St

Japan Center

Richmond

Geary Blvd

Fulton St

Haight-Ashbury

Sutro Heights

Golden Gate Park

Sunset

Golden Gate Bridge. Photograph by Robert A. Rundstrom.

commandant's headquarters (now the Officers' Club) on Moraga Avenue preserves part of the original adobe walls of this earliest San Francisco building.

Tourists often remark how different and natural-looking the Presidio grounds appear in comparison with the rest of the city. The landscape here is entirely of human design, however, complete with curving streets—a rarity in San Francisco—and thousands of eucalyptus and Monterey cypress planted by the U.S. Army in the 1890s to stabilize the dunes underlying this part of the city.

The army is scheduled to relinquish the Presidio in the mid-1990s, when it will likely become a unit of the Golden Gate National Recreation Area. Now, especially on weekends, joggers and bicyclists are more commonly seen than people in uniform.

MARINA DISTRICT

Turn left (north) on Lyon Street and follow it around the *Palace of Fine Arts* and *Exploratorium* (an innovative hands-on science museum, a good stop for families) to its end at Marina Boulevard. The massive Palace of Fine Arts was built in 1915 as part of the Panama–Pacific Exposition, in celebration of the recently completed Panama Canal. The canal strengthened California's economic advantages, but this grand facility was also intended to herald the city's rise from the ashes of the infamous 1906 earthquake and fire. Ironically, the adjacent Marina residential area is prone to more severe earthquake damage than any other residential area in the city.

Turn right (east) on Marina Boulevard. Immediately jog right on Baker Street, immediately left on Jefferson Street, and right onto Divisadero Street. This part of the city is supported only by loose, unconsolidated alluvial sediments that were used to fill in edges of the bay. Solid rock only begins a few blocks farther south. Built in the 1920s, the *Marina District* is the only major residential area built on artificially produced sediments, or fill. During the Loma Prieta Earthquake of October 1989 (7.1 Richter magnitude), the area rolled and shook like nervous gelatin, and the results are still evident in this pricey, waterside community. Look for completely

The Palace of Fine Arts near the Presidio. Photograph by Martin S. Kenzer.

rebuilt lower stories and whole new buildings strangely out of context, such as the site at the southwest corner of Jefferson and Divisadero where the much-televised Marina Fire occurred in 1989.

The Marina's instability is aggravated by 1920s construction techniques. Most of the homes have full basements at street level—note the garage doors and entryways on the first floor. When the fill shook as it did in 1989, the houses rocked, or flexed from side to side like wobbly parallelograms, producing disastrous results throughout the area. Like most San Franciscans, however, residents here are extremely proud of their neighborhood, and they have returned to try again. Solid ground is underwheel again after crossing Alhambra Street.

UNION STREET

Continue south on Divisadero and turn left onto *Union Street*. You will make a loop through one of San Francisco's classic, ritzy, eclectic neighborhoods as you go along Union Street, turn left on Webster, and left again along Filbert Street. Among the homes built after 1906, you will see a few pre-earthquake buildings of the Victorian era and the Vedanta Temple at 2963 Webster, which has been a functioning Hindu temple since 1905. Take your choice of any of the next half-dozen streets to turn left, past Union Street to Green Street. Turn left on Green Street and continue to Van Ness Avenue, where you turn right (south).

An earlier name for this area, "Cow Hollow," has been revived and is used affectionately in the business signs visible along the streets. The prideful use of this bucolic place-name is ironic because Union Street was the first commercial strip to undergo gentrification in San Francisco (1960s), and it is now a popular, upscale shopping and entertainment district, a far cry from its origins as a bovine pastureland. Finally, note the distinctive Holy Trinity Russian Orthodox Cathedral at the corner of Green Street and Van Ness Avenue, a parish in continuous existence since 1859.

VAN NESS AVENUE

Van Ness was always the widest thoroughfare in the city (until Geary Boulevard was widened), a special characteristic with regard to the great fire of 1906. Firefighters held a crucial fire line here, saving the area to the west from destruction. Downtown lay in almost total ruin afterward, and many businesses later relocated to Van Ness or beyond. The area has functioned as an "auto row" and a major hotel and restaurant district for several decades now, reflecting the persistence of the pattern established in 1906. Thousands of dwellings built between 1850 and 1906 survive today, but the vast majority are west of this strip.

PACIFIC HEIGHTS

From Van Ness Avenue, you turn right onto Washington Street and travel west. At Buchanan Street, turn right and then immediately left onto Jackson Street for two blocks to Fillmore Street. You are now in *Pacific Heights,* a persistently wealthy hilltop neighborhood that was shielded from the 1906 fire, and the first of many points affording panoramic views of the city and the Golden Gate.

These hilltop locations, including Russian Hill, Nob Hill, and Telegraph Hill, have retained stable, upper-class residential area in the central city for over a century, and continue to do so—an uncommon phenomenon in urban America.

This is one of the best "house-browsing" sections of the city for the architecturally minded visitor. A number of striking mansions from the 1870s and 1880s are visible here, including sugar baron Rudolph Spreckels's French-style home at 2080 Washington Street (near Octavia Street). Italianate and Queen Anne Victorians abound, many with embellishments carved from California redwood, a common facet of early San Francisco's residential architecture. One garishly painted example presides at 2457 Buchanan Street (corner of Buchanan and Jackson). Perhaps some of these old mansions recall the 1990 movie *Pacific Heights,* in which a young couple buys and restores an old Victorian home only to be terrorized by one of their tenants.

Using Fillmore Street, descend from Pacific Heights and traverse three oddly juxtaposed neighborhoods. Pacific Heights adjoins the Western Addition and the Fillmore District (beginning one block farther west), whose residents are primarily poor African Americans. The landscape changes very quickly again—in a few blocks—as the traveler enters the ethnic, middle-class at Japantown.

NIHONMACHI (JAPANTOWN)

San Francisco's 12,000 Japanese-Americans are distributed throughout the city, but *Japantown* has always been a focus for this community. Go down Fillmore Street to Sacramento Street. Turn left (east) on Sacramento to Octavia Street and continue south on Octavia to Sutter Street. Go west along Sutter Street, turn left on Fillmore Street, and then turn right onto Geary Boulevard. This route takes you past some unique landmarks on Octavia Street: Saint Francis Xavier Church (northwest corner of Pine Street and Octavia), which offers mass in Japanese or English; the Buddhist Church of San Francisco (diagonally opposite Saint Francis Church, and also bilingual); and the National Headquarters of the Buddhist Churches of America next door at 1710 Octavia. Note also the distinctive architecture of the Morningstar Montessori School directly across the street on Octavia. Incidentally, although there are far more Chinese Buddhists worldwide, the form of Buddhism most influential in California, Zen Buddhism, emanates from Japan. Japanese immigrants have come to San Francisco since 1860, but from 1942 to 1945 most were removed from their homes here and placed in internment camps. African Americans moved into Japantown during this period, and are still visible along its fringes today.

As a cultural center for locals and tourists, the big Japanese Trade Center here—between Sutter Street and Post Street and Geary Boulevard—is not as successful as expected. The "Kabuki 8" theaters, and "Denny's" restaurant with its Japanese-inspired motif, for example, indicate the crass commercialism that has developed. However, there is a good bilingual bookstore (Kinokuniya), and the locally renowned Soko Hardware at 1698 Post features hard-to-find Japanese gardening tools and packaged seed.

By leaving Japantown east on Geary Boulevard, travelers pass on the northern edge of an area relatively flat by San Francisco standards. Presiding over this plain is the controversially gargantuan Saint Mary's Cathedral, which draws Catholic San Franciscans of every persuasion from all over the city. Now, the route crosses Van Ness again and heads toward Russian Hill and Downtown.

POLK HOLLOW (POLK GULCH)

A left turn onto Polk Street from Geary brings motorists into a small valley, or hollow, at the western base of Nob Hill. The street narrows, traffic slows in another eclectic neighborhood, albeit one less refined than Union Street. *Polk Hollow* is influenced by the Tenderloin District just to the east, a traditional area trading in prostitution, illegal substances, and various commodities serving prurient interests (for example, O'Farrell Theater with its jungle mural on the corner of O'Farrell Street and Polk, one block south of Geary). Yet, Polk Street boasts some of the finest of San Francisco's more than 4,000 restaurants including the historic

Maye's Oyster House. (Also, note the cable car tracks as you cross California Street.)

Long a nonconformist neighborhood, Polk was one of the first openly homosexual communities in California, but many gays left for the Castro District in the 1970s and this area began to enter a period of relative decline. Now, Asian expansion into the nearby Tenderloin promises to add yet another flavor to the Polk melting pot.

RUSSIAN HILL

Ascend the second major hill, *Russian Hill,* turn right on Washington Street and pass through a small Chinese neighborhood along the way. (A Chinese Kingdom Hall of Jehovah's Witness is half a block left on Hyde Street.) Follow the cable-car tracks of the Powell–Hyde line from Hyde to Leavenworth Street, then turn left onto Leavenworth. All visitors, young and not-so-young, will be fascinated by a visit to the historic Cable Car Barn three blocks farther east on Washington at Mason Street.

The wealthy on Russian Hill were spared in 1906, but development has evolved differently than on Pacific Heights. The core area from Green Street to Chestnut Street is dotted with so many multi-million-dollar condominium towers that the magnificent vistas are often partially or fully eliminated. A forty-foot-height limit has been imposed, but it obviously came far too late for many long-time residents and visitors.

Turn left off Leavenworth onto Green Street, and then again right onto Hyde Street. The corner of *Hyde and Lombard Street* marks the top of one of San Francisco's most famous landmarks, the "crookedest street in the world." Plan to descend through the eight switchbacks and circle back up onto Hyde Street, but realize that a parade of cars accumulates here every single day of the year. Now, put the transmission in low gear again and prepare to descend Hyde Street for one of the most breathtaking urban vistas in the world.

FISHERMAN'S WHARF

The terrain turns to "fill" again at the corner of Beach Street and Hyde, where a local newspaper columnist reportedly introduced Irish coffee into the United States at the famous Buena Vista Cafe.

Fisherman's Wharf and Alcatraz Island in the morning fog. Photograph by Barbara Lerner Kopel.

Ghirardelli Square (an old chocolate factory turned into shops and restaurants), the Cannery, a hotel district, public parking garages, and Fisherman's Wharf are all visible on this route as you go first left on Beach and then turn around to go east on Beach Street to Stockton Street. In short, this is the most popular tourist-oriented section of San Francisco. Great-tasting sourdough bread, freshly steamed crab, and expensive Italian dining are still available, the fishing fleet still sails, and sea lions sunbathe and bark at tourists from the floating sidewalks, but the recent proliferation of T-shirt shops, quick-photo and video stores, gimmicky museums, and gift shop kitsch have sullied much of what this area had to offer. However, the intriguing San Francisco Maritime Museum (on Beach Street, west of Hyde Street) and the Longshoreman's Memorial Union Hall will help to ensure that the orientation of this area stays maritime. If you have a few hours to spare, and plenty of warm clothes, take the boat to Alcatraz Island.

THE SYMBOLISM OF "THE ROCK"

Isla de los Alcatraces (pelicans), *Alcatraz,* was named in 1775 by Don Juan Manuel de Ayala on his first visit to San Francisco Bay. Apparently, the island was left to its avian inhabitants during the Spanish and Mexican periods. No human use is recorded until U.S. Army Engineers arrived in 1853 to erect gun batteries and other fortifications, heralding the island's emerging status as the jagged and scarred sentinel of San Francisco Bay.

Fort Alcatraz began officially as a military reservation in 1859; even then, its primary role was to house prisoners. During the Civil War, acts of treason and mutiny were rewarded with a trip to Fort Alcatraz, where ten-foot-thick walls, a drawbridge, and holding cells excavated out of solid rock evoked comparison with the legendary inhuman qualities of Spanish dungeons. In 1868, the War Department designated the island as a repository for prisoners with long sentences, and during the 1870s and 1880s "troublesome" Indians from Alaska and Arizona Territory were sent here. Spanish prisoners from the Spanish-American War came in 1900, and the worst of our own military criminals were housed here through 1933.

In 1934, Alcatraz became a federal penitentiary designed expressly for the "correction" of only the most difficult and violent of criminals, and those who had escaped elsewhere. Its reputation as the most tightly controlled of federal prisons was based on this mandate, and on its escape-proof history. Al Capone, George "Machine Gun" Kelly, and Clarence Carnes (alias Choctaw Kid)—the only survivor of the bloody 1946 escape attempt—were among the infamous, escape-conscious criminals sent to Alcatraz. In 1962, three men made the only successful escape from Alcatraz, but they were never seen again and presumed to have drowned. The prison finally closed in March 1963 due to its deteriorated condition, and a shift away from isolation as a form of penal correction.

The symbol of The Rock as the home of the incorrigible was overturned in November 1969, when American Indians from the San Francisco Indian Center took possession of the island for nineteen months. The new occupants planned a spiritual center for the teaching and practice of tribal religions, an art colony, a vocational school, and an ecology institute for the nurturing of sacred plants. The original plan called for a positive, pan-Indian home promoting strength, communal spirituality, and the future in a place that had emphasized weakness, personal isolation, and the past. To emphasize their point, Indian leaders also planned to donate a portion of the island to nearby white inhabitants in perpetuity—for as long as the sun shall rise and the grass shall grow—through a "Bureau of Caucasian Affairs" to aid the assimilation of whites to Indian culture.

The 1969 takeover was not part of a contest for land title, nor was it a publicity stunt, but federal agencies construed it as such. The public lost interest once the novelty and media appeal wore off. With momentum subsiding, the original proposal was discarded, and the last occupants were taken off by federal marshals in 1971.

Today, Alcatraz is a part of the Golden Gate National Recreation Area—a unit of the National Park System. Visitors can reach the island by taking a 1-mile ferry ride from Fisherman's Wharf in San Francisco; reservations are recommended during the summer. It seems obvious that the future of Alcatraz is as a recreation facility, yet until quite recently such an idea would have seemed completely out of the The Rock's character.

Turn right on Stockton, and right onto Bay Street. The prime tourist landscape changes between Jones and Mason streets to two blocks of 1950s-era, subsidized, low-income housing that looks ready for demolition. Across the street is a large "Cost Plus" market, the first large-scale importer of Asian products for retail sale in the early 1960s. Such oddly juxtaposed images as these are not unique to San Francisco, but this city seems to offer them with unsurpassed frequency.

NORTH BEACH

Turn left on Jones Street, right on Francisco Street, and you will be crossing the old shoreline again and pass back onto solid ground at the corner of Francisco and Mason. Turn right onto Mason Street: two blocks later, the J. DiMassimo Bocce Ball Courts on the southeast corner of Mason and Lombard signal entry into the traditionally Italian community of *North Beach*. Squeezed between downtown (try not to be distracted by the formidable Transamerica Pyramid just down the street), Chinatown to the south, and imposing Telegraph Hill to the northeast, this tiny, landmark neighborhood is best explored on foot.

Saint Peter's and Paul's Church is where Joe DiMaggio, one of a local fisherman's nine children, was first married in 1929, and where the fishing fleet is still blessed every October. Across Washington Square from the church is the Fior d'Italia restaurant, the oldest Italian dining establishment in San Francisco. Like every other turn-of-the-century business in the city, it operated out of a tent from 1906 to 1907.

North Beach is San Francisco's best offering of a combined ethnic and nonconformist community. Post-World War II North Beach gave birth to the Beat movement, and Italian-American Beatniks were soon known nationwide. Poet Lawrence Ferlinghetti still operates the City Lights Bookstore on Columbus Avenue at Broadway, the most continuously avant-garde, antiestablishment bookstore in the Bay Area.

This neighborhood was not always Italian. French, Basques, Spaniards, and Mexicans were early Gold-Rush-era residents until the turn of the century, when northern Italians and Sicilians became established. San Francisco's fishing and banking businesses were always based in the Italian community, which still retains political power in the city. As elsewhere, Asians are beginning to mix in now, and an enormous amount of Hong Kong money is taking over the banking industry downtown. Look for Chinese- and older Italian-Americans performing t'ai chi exercises in Washington Square daily.

NORTH BEACH/CHINATOWN TRANSITION ZONE

In San Francisco, Naples is just across from Hong Kong: There is literally a one-block transition zone between North Beach and Chinatown. Turn right off Francisco Street onto Mason Street. Go left onto Lombard Street and then immediately right onto Powell Street. On the 1500 block of Powell Street, between Green and Vallejo Streets, John DeLucci's Sheet Metal Works sits adjacent to a Chinese Baptist church and the Cathay Post of the American Legion. Turn right on Stockton Street and note again the juxtaposition of Chinese street signs at the corner of Vallejo and Stockton, the office of an acupuncture association, and a ravioli factory in the 1400 block of Stockton.

CHINATOWN

Stockton Street has always been the cultural and commercial focus for the city's 100,000 Chinese, not the more famous, tourist-oriented Grant Avenue one block east. Most of the schools and churches are here, including the Presbyterian Church in Chinatown that offers services in Mandarin, Cantonese, and English. A five-story low-income housing project in the 1100-block of Stockton interrupts this bustling commercial district with its facade of "government concrete." Across the street at 855 Stockton, the Kong Chow Benevolent Association has been representing the rights and interests of locals since 1854, and two doors down, also note the Chinese-style architecture—using purely Californian construction materials—of Chinese Central High School.

San Francisco's largest ethnic group has also been the most traditionally insular, but recent emigration from Chinatown has reduced Chinese isolation and led to the development of "outliers," or secondary commercial districts in other neighborhoods. Some 35 percent now live in one of these areas out in the Richmond District, especially around Balboa Street and Clement Street. On both Stockton and Grant, the proliferation of T-shirt, electronics, and gift shops has further contributed to the sort of decay witnessed at Fisherman's Wharf. In fact, a majority of Chinese San Franciscans now choose to do their weekly shopping in Oakland's burgeoning Asian community.

Turn right onto Sacramento Street, left onto Powell Street, and immediately left on California Street. You will see St. Mary's Square appears on the right, a park providing a home for Benjamin Bufano's statue of Sun Yat Sen, and for many non-Chinese homeless persons. Turn left on Kearney Street, where in contrast, just three blocks farther on, locals enjoy games of mah jong while businesspersons from the nearby Financial District eat lunch outside at Portsmouth Square. It was here, overlooking San Francisco's harbor just one block away, that the U.S. flag was first raised in California in 1846.

TELEGRAPH HILL AND THE EMBARCADERO

After Kearney Street crosses Jack Kerouac Lane, turn right onto Broadway and pass through the North Beach fringe area of transitory topless clubs and restaurants. *Telegraph Hill* (topped by Coit Tower) looms high on the left, a traditional signaling station overlooking the bay. Turn left on Sansome Street to reach its eastern base. Bay waters once lapped at the bottom of Telegraph Hill where Sansome Street now intersects Greenwich Street. The stouthearted can climb the Greenwich Stairs to Coit Tower.

Turn right off Sansome Street onto Greenwich and immediately right again onto Battery Street. The area around Sansome and Battery Streets was a small warehouse district on the *Embarcadero*, the city's once-thriving waterfront region that declined long ago and has now been renovated. The first Levi Strauss factory was here in 1873. Now, LSU—Levi-Strauss "University"—between Sansome and Battery, and Greenwich and Union Streets, is the $150 million world headquarters of this famous clothing manufacturer. As the first corporation in the city to provide child-care for its employees, Levi-Strauss confirms San Francisco's people-oriented reputation. A city ordinance mandates that 1 percent of new construction space must be allocated to public art, a law with which LSU more than complies. Also, notice the inviting outdoor landscaping in Levi's Plaza on Battery. All this started when a San Francisco man decided to peddle denim from a cart he pushed through Mother Lode mining camps during the Gold Rush.

JACKSON SQUARE (BARBARY COAST)

The area bounded by Pacific, Jackson, Montgomery, and Battery streets was the nefarious "Barbary Coast" of early San Francisco, now known as *Jackson Square*. The 1849 shoreline weaves through this section. Here you could find cheap hotels, saloons, dance halls, brothels, and the sailors and '49ers of that historic era. In this small district, many of the buildings—including some of the earliest the city offers—survived the 1906 fire. Between the 1950s and 1970s, the area was an interior decorator and design wholesaler's district but the businesses moved to SoMa (South of Market). Then Jackson Square became the city's first historic district, raising land values astronomically. This elite district is still home to a few decorators, but fancy restaurants and pricey attorneys' offices dominate. For example, the notorious "King of Torts," Melvin Belli, has his compound at 722–728 Montgomery Street.

THE FINANCIAL DISTRICT

The route now enters the heart of downtown and the *Financial District*. Like North Beach, this area is better appreciated on foot. You can more easily visualize why San Franciscans have become upset by the "Manhattanization" of San Francisco. In addition to what many feel is the destruction of a unique visual skyline, the accelerated wind speeds between the numerous skyscrapers have worsened the local microclimate, already characterized by persistent summer fog.

Drive south down Battery Street, turn right onto Sacramento Street and immediately right up Sansome Street. Go three-quarters of the way around the Transamerica Building: left on Washington Street, left on Montgomery and continue south. Since the late

Sky effects (two lines visible) created by the Transamerica Pyramid. Photograph by Robert A. Rundstrom.

1960s, contractors have designed these buildings so they "wave" under stress, and anchored them deeply in bedrock. The *Transamerica Pyramid,* built in 1972 at 600 Montgomery Street is a classic example of what typically happens during construction in this area. During the excavation, timber and rusty iron emerged at the surface each time another layer of fill was removed. Farther down, the ruins of dozens of Gold-Rush-era ships appeared whose spars and masts had "bobbed" to the surface as the pressure of the fill overburden lessened. Long forgotten were these immigrant ships of 1849 and 1850, whose cargo had hastily departed for the gold fields in the Sierra Nevada. With no one to claim them, the vessels were quickly covered with mud and sand as the city expanded into the water. It is disconcerting nonetheless to found the tallest building in the city among the rotting hulls of an ancient, buried fleet.

Be sure to see the surprising private redwood grove behind the Transamerica building, nurtured solely for employees seeking a peaceful lunchtime repast. Early controversy surrounding the Pyramid's shape has now generally subsided into calm appreciation of a world-renowned landmark.

The intersection at Sansome and Clay Streets marks the former heart of the old printing industry on the West Coast. Printing presses ran twenty-four hours per day on every floor of buildings in this area until the industry was dispersed in the 1960s. Consequently, structures like the one on the northeast corner at 500 Sansome are among the most unshakable in the city.

Finally, although Los Angeles has acquired some of the West Coast banking functions, this area is still the dominant financial center. Bank of America's world headquarters, Wells Fargo Bank, the Federal Reserve Bank at 101 Market Street, and the former Pacific Stock Exchange at 301 Pine Street are all located here. Observe the increasing dominance of Asian banks, including those from Hong Kong, the People's Republic of China, Taiwan, the Philippines, and even Guam.

LOWER MARKET STREET

From Montgomery Street, turn right onto Sutter Street and left on Stockton to go past the elite shopping and hotel area at Union

Square and turn sharply left onto lower Market Street. Market Street is a "break street" between two dissimilar grid patterns, and as such has always been a heavily used retail district. The inconsistent street angles are also the reason traffic inevitably stalls in crossing Market Street.

Lower Market is now enjoying a revival thanks to increasing numbers of Asian and Hispanic merchants. Daily and weekend shoppers here are typically non-white users of the city's mass transit system. All the Latin American consulates—except Brazil's—are located in the Flood Building at 870 Market. Other landmarks include the Hearst Building at Third and Market, the stunning Sheraton-Palace Hotel—built in 1875 and renovated in 1990—and the contrasting concrete and glass Hyatt Regency.

The road crosses onto fill again as you pass First Street (the street name indicates the site of the old shoreline), and the old Ferry Building (now the World Trade Center) appears on the other side of the Justin Herman Plaza at the foot of Market. The city's self-reflective sense of humor about earthquakes is manifest in the plaza's walk-through sculpture composed of 101 concrete cubes, entitled "Ten on the Richter."

Note the statue of Juan Bautista de Anza after turning right from Market onto Steuart Street. The statue was a gift to the city from the state of Sonora, Mexico, and symbolic of the historic and contemporary links between that country and California. Anza is here honored as the city's "founder" in 1776. Jog right on Mission Street, left on Spear Street, go beneath the Bay Bridge, and turn right onto the Embarcadero

The notorious *Embarcadero Freeway* ran from north of the Ferry Building by the Bay Bridge to Broadway along the waterfront here. Some San Franciscans have long argued for the freeway while others vehemently opposed it as an eyesore. Chinatown merchants prized the Embarcadero's Broadway off-ramp that brought tourists directly to Grant Avenue. For many years, completing the Embarcadero Freeway was a heated political topic on which many local elections were based. The 1989 earthquake decided the unfinished freeway's ultimate fate, however, and it is now being dismantled after suffering irreparable fractures.

Thus, San Francisco remains one of the few cities in the United States without a limited-access crosstown highway.

SOUTH BEACH

South Beach is a new place-name. It replaces China Basin, which symbolized the contribution of San Francisco's port to the Asian trade network (there is also an India Basin). The port now moves approximately two percent of West Coast cargo. Because horizontal space is at a premium there is little chance for a significant revival. Indeed, the China Basin and other harbor areas along the Embarcadero are almost entirely filled. Not feeling the same need to link with the past as the residents of Cow Hollow, newcomer baby-boomers here prefer South Beach as a way of aligning themselves with the traditions and hipness of North Beach. Little in the new restaurants, travel agencies, condominiums, and the promenade and yacht harbor refers to anything from the industrial past.

SOUTH OF MARKET

Leave the Embarcadero, proceed northwest on Third Street, and pass under Interstate 80 to enter the South of Market area, the largest neighborhood in the inner city. The name is often abbreviated to *SoMa* to reduce the historical affiliation with gritty, upper Market Street and to acquire the sophistication of New York City's SoHo district. The vernacular of long-time Bay Area residents, South of the Slot, refers to the old cable-car tracks on Market.

Until 1906, the area had a history of housing the most recently arrived working-class immigrants and gained a predominantly Irish flavor as a result. Following the earthquake, SoMa emerged as an area of specialized, small-scale industries with a fairly large "skid row" of seedy hotels and liquor stores. In the 1980s, the area began to change radically for the first time.

SoMa is currently the fastest changing part of San Francisco, with a bewildering jumble of juxtaposed uses. By day, warehouses, small industrial plants, and machine and auto-repair shops operate alongside interior designers, avant-garde art galleries, the increasingly popular factory outlets, and small Filipino-owned businesses. At night, most of the area turns into a multi-ethnic,

Mural advertising condos in the SoMa district on the side of an old flophouse. Old low-rise industrial warehouses and cheap hotels line the streets next to high-rise residences and the new businesses of the service economy. Photograph by Robert A. Rundstrom.

multisexual pleasuring ground of bars, clubs, and restaurants—definitely the "in" section of nighttime San Francisco. These diurnal changes sometimes occur in the same building whereby, for example, a few of the trendiest clubs exist in auto-repair shops.

Turn left off Third Street onto Howard Street (at the Moscone Convention Center), go left at Eleventh Street, and backtrack (left) onto Folsom Street. The strip between Eleventh Street and Sixth Street on Folsom is about as eclectic as possible. In this seven-block stretch, one can find juxtaposed industrial, retail, wholesale, residential, and entertainment functions, in upscale, downscale, gay, or "straight" varieties—and mixtures ad infinitum. For example, Hamburger Mary's at 1528 Folsom has been notable for its leadership in welcoming all sexual preferences. Today, you can

SAN FRANCISCO:
THE CITY THAT WAS

Throughout the nineteenth and well into the twentieth century, San Francisco was the preeminent city in the American West. It was the financial, cultural, intellectual, and social hub west of the Rocky Mountains. San Francisco was a city whose rich ethnic diversity reflected the succeeding waves of immigrants who had been lured there by the promise of better lives: descendants of Chinese railroad builders; Italian and Portuguese fishermen; Russian and Jewish merchants; southern blacks; and Mexican farmworkers and day laborers. It was an enchanting city—full of color and excitement—and it imparted a sensation that great things were happening.

San Francisco has changed significantly during the past quarter century. Its downtown financial district—once the locus of vast wealth, where the "big deal" was closed—is now little more than a collection of glass and concrete boxes that mask the hills and blot out the sunshine. The financial power that was once San Francisco now resides to the south, in Los Angeles, with the transfer of the Pacific Coast Stock Exchange. The turn-of-the-century row-houses that once housed fishermen, longshoremen, and factory workers are now high-priced homes for well-educated, lavishly paid professionals. Cable cars, once the primary means to get about the city's steep hills, have become a tourist trap more suitable to an amusement park.

San Francisco has become a two-tier city, with the affluent and the nonaffluent, the served and the servers, the haves and the have-nots. Gentrification of older neighborhoods has drawn in the wealthy and driven out the poor—a trend that runs against the grain of most urban areas, where the well-to-do move to the suburbs and leave the poor behind in the central city.

By the year 2000, unless the present trend reverses itself, San Francisco will have completed the transition from a city of extreme cultural diversity to one of great sameness, interrupted only by tourists and conventions. San Francisco has enabled Los Angeles to become the new melting pot of the American West, where immigrants flock to seek their fortunes in a dynamic economy. San Francisco has thus become a caricature of the city that had once been the jewel of California and the envy of millions.

still find whatever you are looking for in SoMa. Turn right at Sixth Street and again right onto Harrison Street, which you follow under Highway 101 into the Potrero Hill district.

POTRERO HILL

After passing through the new location of the wholesale fashion, jewelry, and design industry—after its exodus from Jackson Square—turn left onto Seventeenth Street. Behind the Leather Center, another wholesale outlet, one block up DeHaro Street lies the Anchor Brewing Company, famous for its steam brewing process. This is one of the few large-scale (by San Francisco standards) industries, although even Anchor Brewing caters to elite, trendy—and expensive—tastes.

We now swing through the southern ring of residential neighborhoods beginning with *Potrero Hill,* which along with Telegraph Hill, is the only residential hill with an incomplete street grid. Like SoMa, the Potrero provides an excellent opportunity to see contemporary urban processes in action, mainly because it has only recently been rediscovered by other San Franciscans. Also like SoMa, the colorful murals indicate social change, and the Potrero Hill mural at the corner of Seventeenth and Connecticut streets is an excellent example.

Turn right onto Connecticut Street, continue to Twentieth Street. Jog right and then left onto Carolina Street. The city's hilltop social gradient tends to tip south, from the elite on Russian Hill and Pacific Heights to the middle and lower classes on Potrero Hill and Bernal Heights. Its proximity to the shipyards ensured that the Potrero was a blue-collar community of mixed white, African-American, and Hispanic residents. Along with Bernal Heights, Potrero Hill is the only hilltop community not inhabited by an elite, upper class, but the visitor can see that signs of change are evident.

With the port in decline, and moneyed baby-boomers looking for hillside vistas, land values have risen in the Potrero as it has become partially gentrified. The neighborhood you see now is indeed split between a primarily white, upper middle class on the northern slope, with dramatic views of downtown, and a primarily African-American lower class on the southern slope with equally dramatic views of existing port facilities from above a sheer cliff. Freshly renovated homes, iron gates, "Neighborhood Watch" signs, and an abundance of large dogs are sure indicators of white "urban pioneers" on the northern side.

THE MISSION DISTRICT

Turn right to take Twenty-third Street across U.S. Highway 101, which effectively divides the Potrero from the *Mission District*. After crossing Potrero Boulevard and turning left on Hampshire Street and right (west) onto Twenty-fourth Street, the neighborhood immediately changes to a microcosm of Latin America. Although Korean and other Asian merchants have come in recently, Spanish remains the language of choice in Mission businesses, and the dialect is more likely to emanate from Nicaragua, Guatemala, El Salvador, or Peru than from Mexico. There have always been many Hispanics; and the Mission is where most of them are today. The gastronomic implications are significant—this is the place for cheap, good food. This is also one of only a few San Francisco neighborhoods where you will see lots of children.

Two Twenty-fourth Street murals are worth attention: on the walls surrounding a playground in the 2800 block, and on the southeast corner at South Van Ness Avenue. The former speaks to children about Latin American history; the latter trumpets the pursuit of dreams in the "New World." More understated, and consequently more powerful as cultural icons, are the block-length murals on both sides of Balmy Lane that unify walls, fences, and garage doors. Turn right onto Mission Street and note the palm trees and blue tile decorating the sidewalks. This section feels more like an urban shopping district, with its theaters and franchise video, shoe, and drug stores, than the markets on Twenty-fourth.

In truth, the Mission is home to Asians, Africans, whites, gays, and lesbians as well as Hispanics, all segmented into their own sub-neighborhoods. All are united however, by the threat of gentrification expanding from the Castro District to the west; residents do not want to live in another SoMa. They want to keep low-income housing and working-class employment, while confining gentrification to the western Mission. But some blue-collar employers have left, and the Twenty-fourth Street Merchants Association wants to market that strip for more tourist dollars, a process already under way.

THE CASTRO DISTRICT

After crossing Twenty-second Street, Hispanic dominance begins to decline, replaced by white gays and lesbians in the Valencia Street area who have migrated east from the *Castro District.* Turn left onto Eighteenth Street. The first open and organized gay community in the United States begins at the corner of Sanchez and Eighteenth streets. The Castro may be the best example of a traditional nonconformist neighborhood, and has been since the early 1970s. Turn right onto Sanchez, left on Sixteenth Street, and left again on Market Street, and make a sharp right (north) on Castro Street. You will pass the entrance to the Castro Street Station of the Metro system.

Homosexual communities are traditionally marked by increased land values following renovation of whole neighborhoods. Some houses here have been exquisitely refurbished. Male-oriented businesses abound in the Castro, many with striking names—"A Different Light," a bookstore at 489 Castro Street, and "Does Your Mother Know," a card shop at 4079 Eighteenth Street, for

ARE HUMAN VALUES IN THE SAN FRANCISCO BAY AREA STILL "DIFFERENT"?

The Bay Area has always had a unique reputation based on the human values that featured prominently here after World War II. Environmental appreciation and ecological understanding are emphasized, parks and recreation are important, as is development and nurturing of the arts. Art enthusiasts and critics often claim that the Bay Area is one of the last real "bohemias" for arts development in the United States. Galleries, cafes, bookstores, and parks proliferate on both sides of the bay to serve those seeking enlightenment and the means to expression. In short, the Bay Area is often viewed as a healthy, enlightened, humane place to live. First-time visitors will be struck by the number of happy faces—even on foggy days—and the general sense that people are pleased to be living here.

Part of this regional attitude can be explained by the widespread belief among residents that people can make a difference, and that change is not only possible, but that it can be a positive force in everyday life. This attitude plays out in different ways, but travelers can see it in all sorts of signs, schools, community centers, and a variety of retail businesses throughout the region.

It has been traditional here to believe that positive change is best accomplished through the open assertion of individual and neighborhood rights, especially in the face of real or perceived threats of oppression. Monitoring and questioning all formal authority, including police and military forces, is an intrinsic value evident in the Bay Area's urban landscape. A popular automobile bumper-sticker advises, "Question Authority." Other examples can be found in roadside billboards: "Sexual Rights are Human Rights," and "Uhuru:

Keep Drug Task Force Out of the Black Community." Such anti-establishment attitudes are still common, but recent changes suggest that a new value is emerging alongside the old one. Popular support for a free-wheeling, free-thinking society appears to have diminished in recent years as the liberal 1960s and 1970s youth aged and assumed family and business responsibilities. Today's Bay Area youth are more conservative, perhaps in rebellion against the perceived early excesses of their parents.

The 1991 Persian Gulf War provides an excellent example of the emerging tension between these old and new values in the Bay Area, and the relative decline of the anti-authority trademark. In February of that year, the city boards of supervisors in San Francisco, Oakland, and Berkeley declared their cities sanctuaries for anyone facing legal charges for refusing to serve in the military, a measure that would have gone unquestioned a few years ago. Public outcry was immediate and strong. City officials were accused of being unpatriotic, not supporting the troops, and more importantly, being out of step with changes in metropolitan thinking.

The chamber of commerce took out huge advertisements in East Coast editions of *USA Today* and the *Wall Street Journal* to try to stem the anticipated tide of resentment against the city (with consequent loss of tourists' dollars). A vocal, conservative minority has emerged, and respect for authority is on the rise in this traditionally anti-establishment region.

example. Harvey Milk Plaza at the Castro Street Metro Station commemorates the first openly gay member of the county's board of supervisors who, along with Mayor George Moscone, was assassinated in November 1978 by Dan White.

The Names Project at 2363 Market Street testifies to the enormous impact AIDS has had on the Castro and other neighborhoods in the city. This storefront houses the massive quilting project that

View eastward over San Francisco—down Lombard Street toward Coit Tower, on Telegraph Hill. Photograph by Stephen Maikowski.

has been on display nationwide as both a memorial to the dead and a tool for raising national consciousness.

AIDS has been partly responsible for the chain of migration currently rippling through a number of residential neighborhoods here. Recently, a number of "straight" families with children have moved into the Castro, in part, due to the financial toll the disease is exacting from this formerly affluent community. Now, families here want quiet streets at night, so gays and lesbians are moving east into the Mission, raising property values as they do so. In turn, some low-income Hispanics in the Mission are now moving south along Mission Street, to neighborhoods out near the city limits, to escape incoming gentrification.

THE HAIGHT–ASHBURY DISTRICT
Castro Street passes near the hill of Corona Heights Park. Castro changes its name to Divisadero here. Turn left (west) onto Haight

Haight-Ashbury house fronts with elegant gates to keep the homeless off the steps. Photograph by Robert A. Rundstrom.

Street and pass Buena Vista Park. Oddly enough, one of the few places you cannot see easily from the hilltop park at Corona Heights is the *Haight–Ashbury district*, another of San Francisco's famous nonconformist neighborhoods. We enter the Haight through its seedier eastern side beginning at Haight and Central Streets. Note, for example, the graffiti, liquor stores, and decrepit housing mixed in with some renovated Victorians of the late nineteenth century. A few blocks farther, prolific antique stores and vintage clothing shops reflect a focus on a marketable past. In the 1960s, tens of thousands of America's youth migrated here with music, drugs, and peace on their minds. Bands had "houses" then, the most famous of which are the Grateful Dead's, 710 Ashbury Street, and the Jefferson Airplane's, 2400 Fulton Street. Both have been renovated and made appealing for tourists.

The meteoric rise of the hippie era was matched by the speed of its decline in 1969 and 1970. Perhaps the era is best summed up in the motto seen in a local clothing-store window, "Live fast, die young." Yuppies and homosexuals have partially gentrified the neighborhood in the two decades since, but the glossy sheen of the Castro or Cow Hollow is not present here. This is not a pricey shopping district, and the Haight remains one of the few affordable white neighborhoods in the city. Meanwhile, the aura of creative nonconformism lives on at "Bound Together: An Anarchist Collective Bookstore," 1363 Haight, and in the small "rock clubs" that still emphasize home-grown talent, especially "The I-Beam" at 1748 Haight and "Rockin' Robin's" at 1840 Haight.

GOLDEN GATE PARK

Enter *Golden Gate Park* by turning right onto Stanyan Street and left onto Kennedy Drive, which runs westward the length of the park. The appearance of cycling and skating stores on Haight Street signals proximity to Golden Gate Park. Visitors should pick up the park's brochure for detailed information. Major attractions include the outstanding conservatory; the DeYoung Museum—renowned for the Avery Brundage Collection of Asian art; the Japanese Tea Garden, built in 1896—the oldest Japanese-style garden in the United States—the buffalo paddock; and the old

The working "Dutch Windmill" at Golden Gate Park. The cypress trees have been shaped by winds off the ocean. Photograph by Robert A. Rundstrom.

Dutch Windmill before reaching the ocean at the Great Highway where you turn right (north).

SUTRO HEIGHTS

Golden Gate Park is built on sand dunes, and the bluffs to the east overlooking the Great Highway are indeed raised dune surfaces. On the right after crossing Balboa Street, note the "spraycrete" that has been applied in an attempt to prevent erosion on this hillside. On top of that bluff are the ruins of millionaire Adolph Sutro's estate. On the left, the historic Cliff House, a popular rendezvous since 1850, is one of the most famous sites in San Francisco. Elegant dining with unforgettable views are the norm in this living legend now operated by the National Park Service.

Sometimes you can glimpse the seals and sea lions out on *Seal Rock*. Also, walk out to the seaside ruins of the *Sutro Baths*. Erected in 1896, the baths included six swimming pools with a choice of fresh or salt water, and a restaurant, theater, and museum, all enclosed under one enormous glass roof. Up the hill to the right you can see the effects of the incessant wind on the groves of Monterey cypress.

Up the hill and to the left, El Camino Del Mar leads to a large parking lot on *Point Lobos* from which to view the Golden Gate on a clear day. An excellent map explains how the system of lighthouses and fog horns guides contemporary mariners through the frequently fog-bound passage.

RICHMOND DISTRICT

A differently oriented street grid extends from Point Lobos Avenue (the continuation of the Great Highway as it turns east) to Arguello Boulevard, the major artery linking Golden Gate Park and the Presidio. The large *Richmond District*, or simply The Avenues, was build during the 1920s and 1930s and is one of San Francisco's newest residential areas. The houses are the full-basement type like the ones in the Marina, but are less prone to earthquake damage because of the underlying dune and bedrock surface.

Point Lobos Avenue slides into Geary Boulevard about eight blocks inland. Turn right off Geary at Thirty-eighth Avenue and turn left at Balboa. The two Chinese communities centered on Balboa Street between Thirty-eighth and Thirty-fifth avenues, and farther in on Clement Street, are indicative of the Chinese diaspora from the downtown Chinatown area. The westernmost neighborhood, Balboa, is the most recently resettled, whereas secondary migration began to reach Clement Street in the 1960s. Approximately one third of San Francisco's large Chinese community now lives in the Richmond. Follow Balboa east to Twenty-fifth Avenue, turn left and go past Clement Street to turn right on California Avenue.

CALIFORNIA STREET

We cross the city's breadth via California Street. Note that the housing gets systematically older as we near downtown, with the exception of the newer structures on the east side of Van Ness.

Victorian renovations on California Street near Japantown. Photograph by Robert A. Rundstrom.

At Fifth Avenue, California bends gently in preparation for alignment with the downtown grid at Arguello. From there, it is a relatively quick drive toward downtown in non-rush hour traffic. The route passes over the heart of Nob Hill, originally a derogatory reference to its elite "nabob" residents, and home to San Francisco's most famous palatial hotels. California Street then pitches steeply down the eastern side of Nob Hill—with an average grade of 15 percent—as all of downtown lies before you. The street stops at the far eastern end of Market Street, not far from Bryant Street on-ramp to I-80 and the Bay Bridge.

◿ Day Five

WINE COUNTRY

Find Highway 101 and head north, toward the Golden Gate Bridge. After crossing the bridge, take the first exit to the vista point. From this vantage, look back across San Francisco Bay toward the city. This is one of the most appealing perspectives of San Francisco and its environs. You will also note the isolation of Alcatraz Island in the center of the bay, and the swiftness of the currents that surround it. After enjoying this break, return to Highway 101 north and continue up the Waldo Grade. Below you, to the left, are the buildings of the old Fort Baker Military Reservation, with Angel Island beyond. (If you want to take the side trip to Fort Ross, take the Highway 1 exit north.)

After you pass through the rainbow tunnels, portions of the city of Sausalito will be visible below you, to the north. Highway 101 will then drop back almost to sea level, at Richardson Bay. In the bay you see the remnants of the houseboat fleet—long-term residences for some. Across the arm of Richardson Bay are the exclusive communities of Belvedere and Tiburon (Spanish for shark's point).

About 8 miles north of the vista point, the highway crosses Corte Madera (cut wood) Creek. Across the channel, the compounds of San Quentin Prison can be seen. Behind, and east of the prison, the busy arteries of the Richmond–San Rafael Bridge come into view. The city of San Rafael is another 2 miles along the route.

San Rafael is the Marin county seat, the site of Mission San Rafael, and a major, regional economic center. About 2 miles north

San Francisco Bay Area

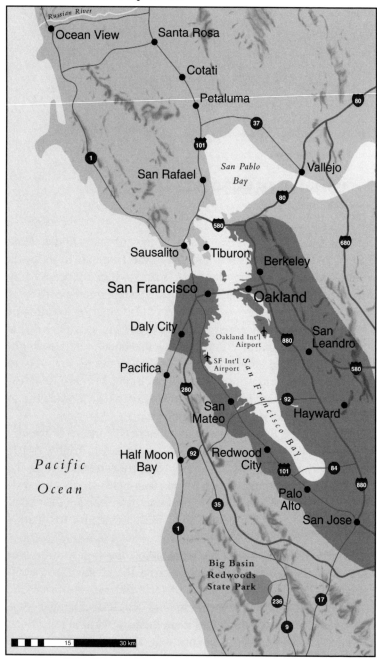

THE RUSSIAN CONNECTION AT FORT ROSS

From 1812 to 1841, one European empire made a little-known but serious attempt to colonize California from the north. The first European settlement on the north coast of California was actually the southernmost extent of the early nineteenth-century Russian empire (with the exception of a brief attempt in Hawaii). Although they eventually failed, the Russians left the most obvious imprint of their effort at what is now Fort Ross State Historical Park in Sonoma County, just north of the Russian River, at the end of a beautiful and leisurely two-hour drive along the coast highway—U.S. Highway 1—north of San Francisco. Here, for a small fee, visitors can see an authentically reconstructed Russian settlement, including its famous Russian Orthodox chapel, built entirely of redwood, and excellent interpretive exhibits.

The Russians had sought a colonial foothold in North America since the beginning of the eighteenth century, and news of Russian designs on northern California reached Mexico City in 1770, spurring Spanish colonial efforts northward. In 1799, the tsar granted monopoly over the northern Pacific coast to the Russian–America Company, for the express purpose of controlling the valuable trade in seal and sea otter skins that had become the object of St. Petersburg's desire.

Five years later, the company had established headquarters in New Archangel (now Sitka, Alaska), and wanted to expand southward. In 1806, Count Nikolai P. Rezanov left New Archangel for San Francisco with the express purpose of securing supplies from the Spanish for the struggling Alaskan colony, and of investigating the sea otter grounds on the northern California coast.

Royal prohibitions against trade with foreigners did not prevent Rezanov or Presidio Commandante Arguello from striking an agreement: Fifteen-year-old Dona Concepción Arguello would become Rezanov's bride, and the Russians would get a supply of wheat and a temporary concession to build Fort Ross on a site 18 miles north of Bodega Bay. Not long after, Rezanov died of injuries suffered in a fall in Siberia while traveling to St. Petersburg to seek permission for the interfaith marriage. (A romantic account of the Rezanov–Arguello episode is immortalized in Bret Harte's poem, "Concepcion Arguello.")

The company pushed on, and in March 1812 a contingent of ninety-five Russians and eighty Aleuts—renowned for their skill in open-sea hunting from their *baidarkas* (two-man kayaks)—landed and established Fort Ross (anglicized from *Rossiia*). From 1812 to 1822, almost 200 Russians and Aleuts mixed with the local Pomo Indians who treatied with the Russians to fend off the inhospitable Spanish.

The remote outpost was mildly successful for a time. At its peak in the 1830s, the company maintained a small port facility at Bodega Bay, an Aleut hunting base on the Farallon Islands 30 miles west of the Golden Gate, and three sizeable ranches within 10 miles of the coast—all in addition to the main settlement at Ross. They produced agricultural goods for the New Archangel colony, harvested and traded sea otter skins, and established regular trade with the Spanish in California.

The Fort Ross settlement was never a legal claim in the eyes of the Spanish and, later, the Mexican government. An opportunity arose in 1834 for Tsar Nicholas I to establish diplomatic relations with the new Republic of Mexico, and to negotiate a legal claim to northern California, but the tsar decided against it.

It may not have mattered, for the Russian empire was grossly overextended by the 1830s. Supply lines by sea or across Eurasia were enormously long, costs had skyrocketed, communication with the central government in St. Petersburg was disastrously infrequent, and, most importantly, Aleut hunters under Russian command had annihilated the sea otter population on the California coast. Moreover, U.S. settlers were beginning to encroach from the east.

In 1841, the Russians left California, selling their land, buildings, livestock, and implements to the entrepreneur John Sutter for $30,000, most of which was never paid. Ironically, within the year, these items would begin to help feed, house, and protect parties of U.S. settlers streaming westward over the Sierra Nevada to Sutter's Fort in the Sacramento Valley.

The importance of this Russian chapter in the story of California lies in its contribution to the diverse origins of the settlement landscape. Although the Russian legacy is minuscule compared with California's Spanish and Mexican heritage, the settlement at Fort Ross, a scattering of place-names, and the descendants of Russian/Aleut/Indian ancestry whose Russian surnames still dot North Coast telephone books are testament to a time when California was the desire of more than one empire. Native Californians remain mindful of their "Pacific connections." On an isolated fog-bound terrace in Sonoma County—facing west—stands another reminder of that heritage.

of the city center, the futuristically designed Marin County Civic Center is visible north of the road. The Civic Center was the scene of the dramatic shootout involving "Soledad Brother" George Jackson and others.

Four miles north of the Civic Center is Hamilton Field, an inactive air force base. It is another 6 miles to Novato, a rapidly developing city that serves as a bedroom community for San Francisco. This region also affords some excellent examples of California oak-grasslands. Continue north on Highway 101. About 4 miles north of Novato you cross the Sonoma County line.

Driving through this region you can still recognize the flavor of the largely agriculturally oriented development of Sonoma County as a supplier to San Francisco. Dairy products, fruit and truck crops, chickens and eggs were once staples of Sonoma economy. Much of this agriculture has been devoured in favor of viticulture. About 4 miles north of the county line is the city of *Petaluma*. This was once a port city, linked to San Pablo Bay by the now-silted Petaluma River. Much of the traffic was devoted to the shipment of produce to San Francisco, a task now rendered by truck fleets. Petaluma still serves as an economic hub for southern Sonoma County, and much evidence of agricultural bounty remains.

North of Petaluma, approximately 10 miles, is the town of *Cotati*. The central hub of Cotati is of interest because of its utilitarian, hexagonal layout. Moreover, there are some attractive and uncommon eateries. It is about another 10 miles to *Santa Rosa*, county seat of Sonoma County (as the result of stealing the county records from the city of Sonoma in the last century). Santa Rosa is the largest city in the county, and, indeed, in much of northern California (note that San Francisco and Sacramento are actually central California cities). Just north of the center of Santa Rosa, still on Highway 101, you will see signs marking the exit for Steele Lane.

Exit at Steele Lane and proceed to the left, under the freeway. Steele Lane soon becomes Guerneville Road, if you follow the main flow of traffic. On the western edge of Santa Rosa, along Guerneville Road, we enter the Russian River Valley and the first vineyard appears to the left. Recently planted in the path of suburban expansion, this vineyard demonstrates the high value of vines in Sonoma County today, and their ability, in some instances, to compete with urban uses. On the other hand, turning right onto Willowside Road, older suburban developments are much in evidence to the right, having replaced previously existing vineyards. Along Willowside Road, and then left along Piner Road, are a number of pre-Wine Revolution vineyards. These older vineyards evince a winescape rapidly disappearing from Sonoma County. Vines are planted relatively close together (often 10 feet by 7 feet), are not trellised, and are all head-trained, spur-pruned; and they are not irrigated.

At the southeast corner of Piner Road and Olivet Road rests an even rarer Sonoma County landscape. Here an old vineyard sprouts a

THE WINE REVOLUTION IN SONOMA COUNTY

In the late 1960s the California wine industry began a series of transformations that today is referred to as the Wine Revolution. Huge new hectarages of vineyard were planted and hundreds of new wineries were built within the next two decades. Wine quality improved enormously as better grape varieties were planted using improved viticulture techniques. Wineries began to employ new technology, European methods, and homespun innovations. The North Coast region of the state, which includes the counties north of San Francisco Bay, soon gained a reputation for producing the greatest number of high-quality wines, although other areas, such as the Central Coast and the Sierra Nevada Foothills have also produced highly acclaimed vintages. The two principal wine-producing counties in the North Coast are Napa and Sonoma, and they are also California's two most famous in the world of wine.

Wine is the leading agricultural product of Sonoma County, ahead of milk, apples, and prunes. At the beginning of the Wine Revolution, however, milk brought in many more dollars, and prunes were more widely planted. In 1967 Sonoma County was planted with some 5,000 hectares of vineyard and more than 6,500 hectares of prunes. Today, the vineyard total surpasses 13,500 hectares and continues to increase, while prunes have shrunk to less than 800 hectares. The new vineyards, not surprisingly, have replaced prune orchards, and also apple orchards, pastureland, and hillside brushland.

The leading grape varieties used to make Sonoma County wine today differ substantially (with one exception) from the leading varieties before the beginning of the Wine Revolution. Carignane, Zinfandel, Petite Sirah, Golden Chasselas, and Alicante Bouschet used to lead the way. Today,

Napa Valley grapevines. Photograph by Paul F. Starrs.

Chardonnay, Cabernet Sauvignon, Zinfandel, Pinot Noir, Merlot, and Sauvignon Blanc carpet the winescape. A partial color change has also accompanied the varietal change. In response to market preferences, grape composition of vineyard plantings has changed from 75 percent red, to 53 percent red.

Prior to the 1970s, most wine made in Sonoma County was not bottled here but rather was shipped out in bulk (in tank trucks and in railroad tank cars) to the Central Valley to be blended with wines from that area. Some twenty-five wineries processed the grapes, but the only one whose name was known was Italian Swiss Colony, which was then one of Sonoma County's top tourist attractions. These days, the wines stay home for the most part, bottled in the 150 or so

Cabernet Sauvignon grapes. Photograph by Paul F. Starrs.

wineries, which may import Central Valley wine for their cheaper products. Many of these wineries are now well-known, both in the United States and abroad. The Italian Swiss Colony winery has been purchased by a large corporation and no longer welcomes visitors.

Both domestic and foreign corporations began to invest in Sonoma County wine in the 1970s, so that today corporate ownership and foreign influence are widespread. Huge French, British, Swiss, and Japanese conglomerates own Sonoma County wineries. Smaller companies or individuals from Germany, Switzerland, France, Australia, and Spain have built or purchased wineries, and American companies (such as Pillsbury, Schlitz, Chevron, and Heublein) have owned wineries or vineyards in Sonoma County. The family-owned

Gallo operation, in the Central Valley, owns Sonoma County's largest vineyard and the county's largest wine-making facility.

Sonoma County's vinous efforts are primarily directed toward the production of still table wine, although important quantities of high-quality sparkling wine (champagne type) are also produced. Dessert wines form only a minuscule part of the picture. The Wine Revolution received its greatest stimulus from the new markets that developed for table wine, because until the late 1960s, California concentrated on higher alcohol, fortified (dessert-type) wines. Those wines comprise just a small proportion of California's offerings today in a market dominated by table wine, especially white table wine.

The huge growth in the wine industry in the 1970s led to a desire for stricter regulation of the areal names put on wine labels. Some producers were using some of the most desirable appellations rather loosely, the better to sell their wines. In 1983, a federal regulation went into effect that said that any area named on a bottle (for example, Sonoma Valley) must have its defined boundaries approved by the government's Bureau of Alcohol, Tobacco, and Firearms. Without such approval, an area name cannot be used on a wine label. The government has approved more than 100 wine appellations, known officially as "viticultural areas," about two thirds of them in California, and a dozen of these either entirely or partially in Sonoma County, which has more appellations than any other county in the country.

With this Wine Country fieldtrip, you traverse parts of three of the most important of the local appellations, the Russian River Valley, Dry Creek Valley, and Alexander Valley. Because these viticultural area are also nested within two larger appellations, Northern Sonoma and the North Coast, and because part of the Russian River Valley sits within Coastal Sonoma, you will also, in effect, be visiting those appellations as well.

few rows of orchard trees interplanted among the vines. This common Mediterranean practice was undoubtedly introduced by Italian immigrants who were a major force in the rural settlement of this part of the county and in building the county's wine industry. Across the street, on the northeast corner of the intersection, sits a vineyard that manifests the latest craze in Sonoma County viticultural practices. Drip irrigation supplements water as needed, a technique introduced to the area in the late 1970s. But more recent still is the double-curtain trellising style, which appeared on the Sonoma scene in the late 1980s. In this case, each row of vines looks like two closely planted rows, but each row actually possesses two separate "curtains" of foliage, which opens the heads of the plants more, thus increasing photosynthesis and producing a larger crop. Some believe that, using this or similar techniques, yields can be augmented without a diminution in quality. These vines are cordon-trained into a "double curtain" and spur-pruned.

Traveling north along Olivet Road, a combination of new and old vineyards is evident. Turn left at River Road, and then immediately right on Slusser Road, where a large old vineyard has recently been uprooted and may already be replanted. A short distance ahead on the left, the estate house of *Sonoma-Cutrer Vineyards* dominates the horizon. The unseen winery is one of the county's most high-tech operations and has become famous for its production of premium Chardonnay wines. At the junction with Mark West Station Road go left and temporarily leave the vineyards. The outstanding feature of this stretch is the *Ocean View Farms* dairy, evidence of Sonoma County's second most important agricultural product. At Trenton-Healdsburg Road, where vines reappear, go right and then quickly left on East Side Road, proceeding to Wohler Road and a right-hand turn. Wohler Road crosses the Russian River and ends at West Side Road. Proceed to the right.

You soon emerge into an area of massive vineyard plantings, which carpet the terraces on either side of the Russian River. This area is relatively cool (often referred to as coastal cool) where Chardonnay and Pinot Noir vines hold sway, and where white varieties such as white Riesling and Gewurztraminer also grow.

TRELLISING

Owners of most of the river-terrace vineyards employ trellising and pruning styles that came into vogue in the late 1960s. Before that time, vines were not trellised. With the arrival of the Wine Revolution, growers began to plant their vines farther apart (12 feet by 8 feet), based on recommendations of researchers at the University of California at Davis. They constructed trellis systems, training the vines along wires (the actual system varying a bit, both in number of wires and in their arrangement). The trellis systems allow for easier maintenance and harvesting, including mechanical harvesting. Two principal training methods were employed, either bilateral cordons (where the arms trained along the wire are permanent wood) with spur pruning, or head-trained vines that are cane pruned (where the arms trained along the vine are one-year-old "canes" that are changed every year). More recently, growers have begun to experiment with a variety of trellising systems, including different kinds of double curtains and also pergola arrangements.

The cool summer climate results from nightly inflow of fog-laden marine air, which often presents a low cloud cover that may not burn off until midday or later. The terrace soils are productive, and crop loads must be limited to avoid overabundant yields that would result in lower-quality wines. The lower terrace areas, particularly on the other side of the river (which generally cannot be seen here), are subject to winter flooding. This is not necessarily a problem, but in areas of poor drainage (in particularly wet years) the vines suffer after they bud in spring. Older vineyards appear to the left on the hillsides. Before the Wine Revolution, most of the lower terrace land was planted with prunes. Several wineries are located along this road, including Davis Bynum, Rochioli, Hop

Filling wine casks at Robert Mondavi winery. Photograph by Paul F. Starrs.

Kiln (a spectacular edifice on the right and sign of an earlier agricultural industry in these parts—hops for beer, which disappeared in the early 1950s), and Mill Creek.

After several miles we leave East Side Road for West Dry Creek Road. We enter a separate viticultural area, the Dry Creek Valley, which warms noticeably from south to north during the summer. The upper two thirds of the valley are coastal warm climatically, beyond the usual fog-bank intrusions that penetrate the Russian River Valley. This is the land par excellence of the Zinfandel grape (for big, red Zinfandel wines), but Cabernet Sauvignon, Sauvignon Blanc, and other varieties do well here, too.

Both of the principal means of frost protection used in California can be seen along this road. Particularly on or near the valley floor, without some means of providing warmth, cold-air drainage on nippy spring nights after bud break can cause havoc. (Note that winter freezes are not a problem in California; temperatures do not drop low enough to freeze vines.) The propellers occasionally visible above the vineyards are wind machines, which, in combination with orchard heaters (burning diesel fuel), mix warm air over the vines. This is a single-purpose system that may be effective to temperatures as low as −2 degrees Celsius. More common in Sonoma County is a network of overhead rainbird-type sprinklers, referred to as "permanent set." This means of frost protection depends upon the heat released on the new leaf surfaces as water freezes (the heat of fusion) to protect the green material. This system is effective to −4.5 degrees Celsius, but is much more costly to construct than the wind-machine–orchard-heater combination. It offers the advantage of being a multiple-use operation, however, since the sprinklers can be employed for irrigation or for distribution of pesticides.

From West Dry Creek Road, go right on Lambert Bridge Road (having just passed *Lambert Bridge Winery* on the left). At the junction of these two roads, on the west side, a new, terraced vineyard hovers above. The valley floor is now completely clothed in vines, which has led to the siting of more and more vineyards on the slopes above the floor. Terracing, a rare sight in Sonoma County before 1970, is becoming more and more commonplace as those wishing to plant additional vines are forced to seek steeper and steeper hillside plots. Ahead, a low ridge dominates in the near distance, separating Dry Creek Valley from the Alexander Valley.

The road crosses Dry Creek, which is dammed at the northern end of the valley, forming a substantial impoundment known as Lake Sonoma. Across the bridge, *Robert Stemmler Winery* occupies a spot on the left-hand side of the road; the larger *Dry Creek Vineyards* is in view on the right. Reaching West Side Road, go right and when you see a small sign on the left that says *"Frei Brothers Winery,"* begin to look off the road in the distance to the left for the immense stainless-steel tank farm that is part of this winery. Although still referred to by the old Frei Brothers name, this operation has belonged to the Gallos for some time and is the Sonoma County focus of their extensive wine empire.

Quickly go left on Lytton Springs Road, which traverses the low ridge between the two valleys, pass Lytton Springs Winery, and emerge on the floor of the Alexander Valley. Head north on Highway 101 (the freeway), which offers excellent views of the valley to the right. This valley, like those of the Russian River and Dry Creek, bears one vineyard after another. The principal water course through the valley, from south to north, is the Russian River, but the structural valley it traverses here is named after an early settler. Most of the Alexander Valley is also coastal warm, though its southern portions are coastal cool. Cabernet Sauvignon and Chardonnay reign supreme here, but Merlot, Sauvignon Blanc, Zinfandel, and other varieties occupy plots on the floor and adjacent hillsides of this largest of Sonoma County valleys.

Take the first exit, Independence Lane, and proceed to the left under the freeway, where you pass under the arch that identifies the winery ahead as Chateau Sovereign. The large, architecturally impressive structure employs hop-kiln forms. This edifice was built by Pillsbury, was later owned by a group of 250 or so growers, and now belongs to the Swiss giant Nestlé, which has a growing California wine operation that also includes Beringer Vineyards in the Napa Valley. Return to San Francisco via Highway 101 south.

Oakland and the East Bay

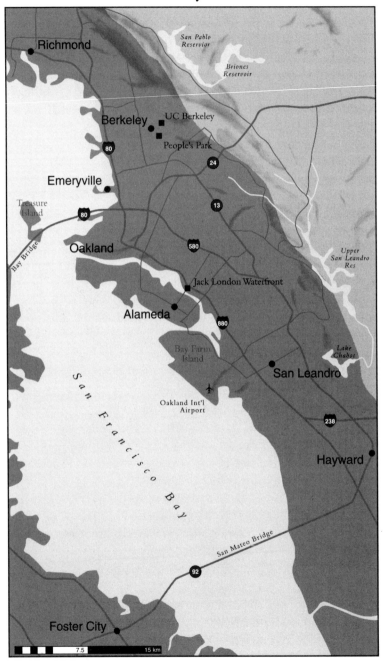

Richmond

San Pablo
Reservoir

Briones
Reservoir

Berkeley   UC Berkeley

80

People's Park

24

Emeryville

13

Treasure
Island

80

Upper
San Leandro
Res

Oakland

Bay Bridge

580

Jack London Waterfront

Alameda

880

San
Lake
Chabot

Bay Farm
Island

San Leandro

F
r
a
n
c
i
s
c
o

238

Oakland Int'l
Airport

Hayward

B
a
y

San Mateo Bridge

92

Foster City

7.5   15 km

◸ *Day Six*

EAST BAY TO THE FOOTHILLS

Now the route leaves San Francisco for the East Bay communities, then the agricultural valleys, and eastward to the foothills of the great Sierra Nevada mountain range. Travel along Fifth Street, south of Market Street, toward Interstate Highway 80 (I-80). The parking lots, small and inexpensive restaurants emphasizing weekday lunches, small groceries, and new retail department stores along this Fifth Street strip indicate the current transitory character of SoMa. The old San Francisco Mint, now a public museum (hours: 10:00 a.m. to 4:00 p.m., Monday through Friday), is on the right soon after Market. On the left, after crossing Howard Street, is the Yerba Buena Square–Burlington Coat Factory complex, now the largest and most expensive multiple-use development in the city. Note also the colorful mural paying homage to the Golden Gate on the building to the right.

Make a sharp left at the Bryant Street on-ramp for I-80 and the Bay Bridge. Unfortunately, the view is obstructed on the lower deck of the bridge, but you can catch occasional glimpses of downtown San Francisco to the left and the South Beach section to the right. The tunnel through Yerba Buena Island occupies the crossing's midpoint, and the bridge changes from the graceful suspension design to the angular cantilever section.

Attentive travelers can identify the section of overhead bridge roadway that, during the Loma Prieta Earthquake in October 1989, slid off its supporting plates and slammed onto the lower deck road surface. The noticeable clues include a change in the "feel" of the paving underwheel, and a slightly newer-looking coat of gray paint. Look for these indicators just before the turn (traveling eastward) in the cantilever span, approximately 1.3 miles from the tunnel exit.

The Port of Oakland

While descending the last section of the Bay Bridge, the justifiably dominant view of the East Bay is of Oakland's huge military and container cargo port off to the right. The Oakland Army Base, and U.S. Naval Supply Center and Naval Air Station, in addition to the civilian commercial port facilities, are easily identified by the giant cargo cranes, and the rail and trucking yards spread at their feet.

As a transportation hub, Oakland offers at least four advantages over San Francisco: its location on the mainland side of the bay as the century-old western railhead of the transcontinental railroad; large expanses of relatively flat terrain enhanced by extensive areas of "fill"; visionary port leadership; and a stable history of labor–management relations. Oakland "scooped" San Francisco by being the first on the West Coast to develop multi-modal container shipping. With numerous trucking and shipping companies already headquartered there, by 1970, Oakland had easily surpassed its cross-bay rival in volume handled. San Francisco finally opened a small container terminal in the early 1980s, too late to garner much business. Today, Oakland handles 90 percent of Bay Area container cargo, and is ranked from second to fifth in the nation, depending on the means by which port business is measured. Its major West Coast competitor in the United States is Long Beach, and secondarily, Seattle. All three are in heated competition with Vancouver, Canada, for the burgeoning trans-Pacific trade.

Emeryville

The Powell Street exit off I-80 east brings visitors into the heart of *Emeryville*, at one time a waterfront city of strictly heavy industry,

DIVERSITY IN THE TWO CITIES BY THE BAY

Travelers can readily recognize the same important physical, economic, and social differences between Oakland and San Francisco that residents know and frequently argue about. San Francisco's history as guardian of the Golden Gate and the best natural harbor in California, and as depot for the Gold Rush, was determined by its unique location as the thumbnail on the thumb of a peninsula. Such a site ensured a temperate but chilly climate on the seasonally fog-bound coast. It also precluded future expansion, and the jumble of hills, hollows, and sand dunes on which the city was built have always made it dependent on a productive hinterland. In contrast, Oakland spreads across a large, uniformly even alluvial plain on the mainland side of the bay, a prime location for the diverse agricultural production that characterized its early existence. The lengthy bay waterfront to the west was originally quite shallow, but dredging allowed easy transport of products to the coast and beyond. Although both cities are equally susceptible to earthquakes, San Francisco is more widely known for them. But water-absorbing serpentine clays embedded in the long ridge of hills to the east make Oakland far more prone to landslides, which often occur independent of seismic activity.

The cities' modern economic functions are more divergent and complementary than competitive. San Francisco dominates in administrative, banking, diplomatic, and other service functions. It is the headquarters of numerous corporations, international banks, and many of the West Coast's foreign consulates. Yet, San Francisco would wither without entertainment and tourist revenue. As a case in point, after the 1989 Loma Prieta earthquake, the crucial issue was not whether the bank buildings were safe, but whether the cable-car system was intact. The city's glamour and European "feel" come at a high cost. Fewer manufacturing jobs exist here than in any other U.S. city of comparable size.

The blue-collar domestic and international functions are located in Oakland. Its dominant role as a transportation hub started in 1869 when it became the terminus of the transcontinental railroad. Some rail cars were ferried across the bay, but most stayed on the mainland side. Now, more than 1,000 trucking companies and dozens of transoceanic shipping companies have offices here. For example, American President Lines recently completed a large, new skyscraper downtown on Broadway. The Port of Oakland handles virtually all the Bay area's non-mineral cargo (oil tankers operate out of Richmond), and the docks are also home to sprawling military facilities. As the Bay Area continued to grow, planners knew it made good sense to design the BART subway system with the main hub in Oakland.

Their economic roles undergird the popular images of both cities. San Francisco symbolizes fun, glamour, and excitement; Oakland is viewed as a mundane, gritty backwater with a rough past. Two anti-authority, violence-prone organizations born in Oakland in the 1960s—the Hell's Angels motorcycle club and the Black Panther party—certainly contributed to the tough reputation, and they still maintain a presence in the city. Gertrude Stein's oft-cited complaint that in Oakland "There is no there there," still stings Oaklanders, although it is now believed she was referring to the more personal loss of her Oakland girlhood.

The same difference is also manifest in professional sports. The Oakland Raiders were a tough, intimidating, do-anything-to-win professional football team of the 1960s and 1970s. Management and coaches took great pride in the number of misfits and castoffs they could mold into winners, and the team was either reviled or applauded nationwide. Today's Oakland A's baseball team includes a group of muscular homerun-hitters, the "Bash Brothers," who enjoy pounding their forearms into each other as a form of bonding. The San Francisco '49ers (football) and Giants (baseball) have always played a more glamorous and ethereal type of game. 28 May 1991, the Giants' owner invited six Tibetan monks to Candlestick Park for a game against the powerful Cincinnati Reds. They blessed the baseball team from their box seats along the third-base line while management stood by, hoping the prayers would reverse a dismal early-season start. It would never work in Oakland, but the Giants won, 6 to 2.

Both cities are proud of their differing social and cultural geography. Strong community identities emanate from a tendency for San Francisco's diverse ethnic population to cluster in mutually exclusive neighborhoods. Oakland's neighborhoods are typically more integrated. In fact, Oakland's new motto, "The Most Integrated City," is not just typical boosterism. Recent census data show that only 2 percent of African-American residents live in wholly African-American neighborhoods, and only 7 percent of all others live on streets with no African-American neighbors. Oakland's substantial African-American middle class has also entered business and government with great success, whereas San Francisco's African-American population remains largely isolated as an underclass in the Western Addition and Hunter's Point.

San Franciscans are rightfully proud of the personal freedom of expression inherent in the avant-garde singles lifestyles found throughout their city, and they have traditionally looked on Oakland as an inferior locale offering little excitement, intellectual activity, or culture. Meanwhile, Oaklanders joke that the only basis for San Francisco's superiority complex is that drivers must pay to get into the city when crossing any of the bridges. They also counter by emphasizing their family-oriented, middle-class neighborhoods and diverse job opportunities.

Of course, the two cities are complements more than competitors, and together with other Bay Area cities they form one of the most diverse urban metropolises in the United States. The East Bay may claim the last laugh, nevertheless. The top four brands of San Francisco's famous sourdough bread—Boudin, Toscana, Colombo, and Parisian—are all made by the San Francisco French Bread Company, the aroma from which can be inhaled by anyone driving within a few blocks of their location at 580 Julie Ann Way—in Oakland.

especially steel fabrication. The Powell Street overpass provides a view of the still-viable industrial area, including the Southern Pacific railroad tracks that come up from Los Angeles.

Large portions of Emeryville have undergone dramatic change. The block-sized Westinghouse motor-repair factory to the left of the overpass, now abandoned, is indicative of the area's growing departure from its industrial past. Turn left onto Hollis Street and witness a "California solution" for turning aged industrial zones into new economic landscapes.

Hollis Street, from here across Ashby Avenue to Carleton Street in Berkeley, is a district of heavy and light industry—including some "environmental" companies—mixed with retail stores and outlets, restaurants, import–export companies, an artists' colony, and some residences. Like the larger SoMa district in San Francisco, Emeryville has traded in its industrial heritage and become an "in" place to eat, drink, live, work, or shop, but without the nightlife across the bay.

The *Hollis Street Complex* and *Heritage Square* are excellent examples of the adaptive reuse of old warehouses. First-floor shops, galleries, and restaurants lie beneath spacious artists' lofts upstairs. These buildings were designed for creative professionals who can no longer afford to live or work in San Francisco, or who find the East Bay more stimulating.

The Hollis strip is approximately a mile long, and worth driving twice. The traveler will notice larger plants such as Ryerson Steel, American Transit Supply, Grove Valve, McGrath Steel, and the massive PQ Corporation mingled with hot-tub dealers, coffee importers, the National Holistic Institute, a lumberyard specializing in expensive tropical woods and redwood, and the nationally renowned Whole Earth Access outlet store, a classic "enviro-yuppie" landmark.

Berkeley

The Hollis Street "phenomenon" continues on the Berkeley side of Ashby Avenue. Noteworthy here are: "Juan's Place," a fine Mexican restaurant on the corner of Ninth Street and Carleton; the offices of Fantasy Records which boosted its first local band— Creedence Clearwater Revival—to national fame in 1969; and Parker Plaza, another adaptively reused complex of artists' lofts above a cafe and other shops.

Hollis Street crosses Dwight Way and jogs left to become Sixth Street. Dwight Way marks your entry into a mixed low to middle-income residential area. Here, whites, African Americans, and Hispanics are interspersed in typical Berkeley residential fashion. The houses are mixed in age too; most were constructed between 1910 and 1960. The shingle-house on the right at 2013 Ninth—the first of several on this route—is an informal style for which Berkeley is well-known, a design made famous by architect Julia Morgan. Also notice the iron gates and barred windows in this neighborhood, suggesting a self-protective atmosphere, whereas the shady sycamore trees, clean streets, and bicycle lane encourage a contrastingly benign view of Ninth Street.

Continue on Sixth Street to University Avenue and turn right (east). University Avenue, a major retail business district, connects the bay waterfront and I-80 with the prestigious *University of California* and associated residential areas in the foothills to the east. Recent immigrants from India have settled here among primarily African-American residents, and the number of restaurants, fabric stores, and other recent Indian businesses has enhanced the ethnic diversity along this previously more homogeneous avenue.

The median strip on University Avenue, like the ones on Sacramento Street and Shattuck Avenue, represents the former rights-of-way of the old interurban streetcar system that connected East Bay cities.

Proximity to the university community is evident after crossing Sacramento Street. Note, for example, the "Living Foods" grocery store on the left where "naturally raised meats" and "organically grown fruits and vegetables" can be purchased. The number of motels, restaurants, and video/audio stores begins to increase, and the street signs switch to "Cal's" colors, blue and gold. Also, sidewalk access for people in wheelchairs becomes evident at street intersections (Berkeley is the center of the Disability Rights Movement). The university is now visible straight ahead, with the Campanile and Evans Hall dominating the skyline.

The International Buddhist Institute and "Sushi California" restaurant, featuring a Westernized Far Eastern cuisine that has now swept the country eastward from California, are on the left immediately after turning right onto Martin Luther King Way (formerly Grove Street). The route winds through downtown Berkeley past the earthquake-damaged City Hall on King, and the park in front that has been adopted by homeless persons. Turn left onto Allston Way, go one block to turn left on Milvia Street, and right on Center Street. Berkeley's central business district emerges after a right turn onto Shattuck Avenue. A BART metro station on this corner provides easy access to unique shopping and the edge of the Cal campus just one block away. Note the college-oriented services on Shattuck, including bohemian cafes and a number of fine, old theaters. Turn left onto Dwight Way (east, up hill).

Berkeleyans are as fiercely proud and protective of their neighborhoods as are people throughout the Bay Area. The concrete pylons obstructing traffic on many of Dwight Way's cross-streets—Fulton Street and Dana Street, for example—are uniquely Berkeley's, however. Although "softened" as planter boxes, the pylons embody the common attitude that in Berkeley, neighborhoods and people matter more than traffic and cars.

Infamous *People's Park* appears on the left after crossing Telegraph Avenue. It is very difficult to predict what visitors will find at this mercurial site. Traditionally, People's Park has doubled as a focal point for political activism and as an overnight campground for the homeless. In 1991, the university built a volleyball court on the site, touching off two days of demonstrations and violent confrontations between local citizens and police. Local residents and tourists enjoy shopping in the unusual stores nearby.

Note the unusually informal *First Church of Christ Scientist,* designed by Bernard Maybeck, on the northeast corner of Dwight and Bowditch Street. The building's bungalow-style roofline and exposed redwood timbers, coupled with the poured-concrete walls, the redwood trees shading it, and the hanging plants imported from Asia mark it as a uniquely Berkeleyan church.

Turn left on Bowditch Street and left down Haste Street. Turn right off Haste Street onto Telegraph. The mural on the wall of the Amoeba Music Store at the corner recounts recent city history in a manner endemic to this south-campus neighborhood. Bookstores and other student-oriented businesses cluster in this business district. Although remnants of its more radical past still exist at places like "Annapurna" and "Rasputin's Records," and in the windowless facade of the Bank of America on the corner of Durant Street, this former "Bohemia of the West Coast" is increasingly invaded by more conservative, upscale businesses reflecting both the homogenizing effects of the baby-boom generation and the wealthier student population inhabiting universities these days. Local coffeehouses have become franchised fast-food outlets or fashionable pubs. Franchised clothing boutiques found in any city have replaced many of the unique second-hand shops. "Head shops" are gone or "underground," and customers cannot buy a beer at La Val's Tavern now without ordering a minimum amount of food first. Even the wares peddled by the street merchants look increasingly refined. There was a scandal a few years ago when it was discovered that an East Bay department store was surreptitiously selling goods here on weekends. Although much of the "authentic" flavor of South Berkeley has been altered, this is still a wonderful

FROM PEOPLE'S PARK TO PEOPLE'S BOUTIQUE

The *Almanac of American Politics* lists Representative Ronald V. Dellums as the most radically liberal congressman in the United States. Perhaps this comes as no surprise, given that the district he has served for more than two decades comprises North Oakland and Berkeley. After all, Berkeley holds the distinction of being the only part of the United States to recognize the Communist government of Nicaraguan Daniel Ortega in the 1970s. Berkeley was also the first city to decriminalize marijuana possession and use (less than an ounce) and was the kind of place where, in the 1960s and early 1970s, it would not have been unusual to spot local cops enjoying a quick "toke" among locals on Telegraph Avenue ("The Av").

Times have changed, and the baby-boomer's homogenization of America has affected Berkeley, too. Although it is still an unusual place, The Av and other areas around the University of California campus increasingly look like the rest of the United States. A case in point is the infamous People's Park.

People's Park is not indicated on most street maps because it is not officially recognized as a park. Indeed, it is undoubtedly like no other park, with the possible exception of Hyde Park in London. The area bounded by Dwight Way, Bowditch Street, Haste Street, and Telegraph Avenue is actually owned by the University of California, however, a key factor in the symbolism of the place.

The university razed a number of decrepit buildings on the newly acquired site in 1967, but competing proposals and lack of funding stalled plans for a gymnasium, student dormitories, classrooms, and ball fields, all proposed at one time or another. Meanwhile, many of Berkeley's "underground" population took up residence on the unused property.

People's Park really began in April 1969, as the result of a conscious effort by university instructors, students, hippies, and political activists to create an alternative "habitat" in Berkeley. People were living year-round on the site in tents, parked automobiles, or under the stars. They were also nurturing gardens, using and dealing drugs, hosting musical concerts in the middle of the night, panhandling on The Av, and generally enjoying nationwide notoriety as perhaps the most recalcitrant of nonconformist places in the country. Indeed in the 1960s Berkeley became a mecca for an experimental youth subculture, down-and-out derelicts, and political activists from all over the United States. People's Park—and its neighbor three blocks south, Derby Street Park (renamed Ho Chi Minh Park in the early 1970s to reflect local political alliances)—became national focal points for Vietnam-era protests and subsequent radical political activity of all kinds.

A riot ensued when the university erected an eight-foot chain-link fence on the site on 15 May 1969. Governor Ronald Reagan placed the entire city under curfew for the next two weeks, and the National Guard patrolled the streets until calm prevailed again. In 1972, the fence was removed and residents resumed their earlier activities. For many years thereafter, officials of the traditionally liberal-minded city and university generally ignored activities in or near the park while they wrestled over its optimum use.

Patty Hearst, heiress to the Hearst estate, was kidnapped in 1974 from her boyfriend's home just three blocks away; and, in the hands of Donald DeFreeze (alias Field Marshal Cinque) and others of the Symbionese Liberation Army, she went on to become a radicalized bank robber purportedly trying to help feed the poor. That this bizarre event developed in the shadow of People's Park and The Av seemed, at the time, quite normal.

In May 1989, a Catholic relief organization opened The People's Cafe to distribute breakfasts to the homeless who lived there. Hare Krishnas served free gourmet dinners, including day-old *haute cuisine* featuring curried vegetables and fresh-baked whole-grain breads. But the mood on campus and in the city was more conservative by this time, and the groups were evicted in March 1990. After all, such activity was against university regulations, although it would have been difficult to count the number of illegal activities at People's Park at any given moment in the years since 1972. The Av was losing more and more long-time retail business after years of difficulty with panhandlers and drug dealers. Also, downscale shops were not as profitable as upscale businesses catering to new, wealthier students and local baby-boomer graduates. Boutique-type shops were waiting in the wings, as elsewhere in U.S. cities, but their conservative, profit-conscious owners were averse to opening here until something was done about "them."

In February 1990, the university and city council jointly unveiled plans to clean up the park, a move seen by the more refined elements of the politically well-organized homeless population as a not-so-subtle pretext for eviction. The diminished number of radicals considered it the political equivalent of a Nazi blitzkrieg. In fact, the plan did include an all-night curfew from 10:00 p.m. to 6:00 a.m., which amounted to eviction. As people were removed, some forcibly, a "copwatch" formed—a Bay Area tradition—to monitor alleged police misconduct on the site.

Events on the site accelerated dramatically in 1991. In mid-July, the Berkeley City Council outlawed sleeping outdoors, and on 31 July the university began construction of a $100,000 volleyball court made with expensive redwood siding. The court occupies just one eighth of the park, but the city knew what was coming next, and city sanitation workers began removing garbage cans and other large objects from the area on the same day. A repeat of the 1969 riots occurred later that evening. Two days of rioting, looting, and random violence led to numerous injuries among police and demonstrators, and 104 arrests were made by twelve different law enforcement agencies. But the volleyball court remains, albeit one of the most heavily guarded ones in existence.

Now, People's Park attracts volleyball players along with political activists, local residents, the curious, and a few persistent homeless. However, Jon Reed, one of the 1969 founders of the park, observed in the *Oakland Tribune* (2 August 1991): "Twenty-two years have passed and nothing much has changed. They'll be bogged down in the park for a long time like the U.S. was in Vietnam." Meanwhile, the "yuppification" of The Av is finally under way.

place to shop, walk, or just absorb the ambience. Street poets and musicians still wander looking for a willing ear.

The University of California's main entrance at *Sproul Plaza* and *Sather Gate* awaits pedestrian visitors at Bancroft Way. You can drive onto the campus either through the entrances on the east or west sides (but do not expect to park on campus). The Student Union and Zellerbach Auditorium are immediately on the right after turning left on Bancroft Way from Telegraph, across from the University Press Bookstore near Dana Street.

Turn left on Dana and go to Dwight Way. Turn left up Dwight Way and right at Telegraph Avenue. Lower Telegraph in Berkeley follows a former streetcar line toward downtown Oakland. This area is now an office and health-care center serving much of this part of the East Bay. On the corner at Ashby is the supermarket-sized Whole Foods Market, and a few blocks farther is the Berkeley Holistic Health Clinic, both evincing the emphasis on alternative health and nutrition in this part of the Bay area.

Oakland

When you cross the Oakland limits at Telegraph and Sixty-sixth Street, you will readily notice the change from Berkeley. Telegraph is a commercial strip from beginning to end, but it functions primarily as an anchor for local business in this older part of Oakland, providing an interesting southwest transect from the north end of the city to downtown.

Attention to the subtleties along the Telegraph strip can be rewarding. For example, most travelers may not notice the unobtrusive blue and white facade of the business on the right side of the street, just past Sixty-sixth. The lone image of a white stallion, without words or a clear indication that this is indeed a commercial establishment, indicates that this building is probably a "gay bar." Further evidence of a homosexual presence in North Oakland is the lesbian bookstore, "Mama Bears," across the street.

African Americans compose just over half the population of Oakland, primarily due to a massive migration from the South, especially Mississippi, between 1940 and 1960 when blue-collar jobs were available here. Although not consistently dominant along Telegraph, businesses definitely reflect the African-American presence here. Rib joints and billboards catering to select viewers are among the more obvious elements of this ethnic landscape. For example, the "Okla [Oklahoma] Hickory BBQ" in the 1930s storefront on the corner at Fifty-ninth Street is a mainstay of the neighborhood. A mural on the side of an old house converted into the Telegraph Baptist Community Church near Fifty-fifth Street, and the "Soul Brother's Kitchen Cafe," renowned among Oaklanders for its outstanding ribs, are further indications of local ethnicity.

The heart of the *Temescal District* is at Fifty-first Street. This was a predominantly Italian neighborhood for a long time, but recent immigrants from East Africa—including more than 1,000 Eritreans escaping the civil war in Ethiopia—have changed the flavor of Temescal. Ethnic juxtapositions like those seen here are typically Californian, and the observant visitor can outline local history in short order, identifying newcomers and established residents alike by observing the names and ages of businesses, buildings, and streets. For example, note in the immediate vicinity: the Aikido Institute, the new Asmara (the Eritrean capital) Restaurant at 5020 Telegraph, a hardware store in an old (1869) Italianate Victorian building, the Genova Delicatessen—visited by gastronomes from throughout the city, and the National Office of the African People's Solidarity Committee (featuring "Uhuru Pies"). Moreover, this "hopscotch" pattern continues beyond Temescal. The Cafe Eritrea d'Afrique at 4069 is just down the street from the older-looking Bertola's Italian Restaurant, and Lucca's Deli at 3838 Telegraph precedes a Pentecostal church, the First African Methodist Episcopal Church, the newer-looking Korean Community Center of the East Bay, and Everett and Jones BBQ near West MacArthur Boulevard. Even Vicente Way, just off Telegraph near Fifty-fifth Street, carries history in its name; in 1836, Vicente Peralta built his adobe on his brothers' rancho land grant near here (where the Chevron service station abuts the freeway overpass).

An old *mortuary district* is next on Telegraph after the I-580 overpass, circumstantially located only a few blocks west of "Pill Hill," the major medical center in Oakland. Most of the mortuaries have moved elsewhere now, but the unique buildings have been adaptively reused. Both the sports-care medical clinic and the athletic club on the right side of the street both occupy old mortuaries. Grant Miller Mortuary at 2850 Telegraph is the last one, and has adapted to ethnic changes by appealing to an increasingly Asian clientele.

DOWNTOWN

The Sears store on Telegraph Avenue at Twenty-seventh Street marks the beginning of downtown Oakland, a good example of an American downtown waiting for something to happen. Note the numerous vacant buildings, bars, "adult videos," and other transient businesses in this north end. The "BUS" furniture store on the corner at Twentieth Street, which was indeed a bus terminal, is indicative of the impermanent retail strength here. The Emporium department store occupying the block to the left was closed by the 1989 Loma Prieta Earthquake, and reopened in 1990 after expensive repairs, including the elimination of all upper-story windows. Many buildings in the downtown area remain closed, however, either due to economic decline or the 1989 quake.

The historic *Fox Oakland Theater*—a classic 1928 art deco movie palace inspired by a temple in northern India—has been closed for many years awaiting renovation. The Fox is a national historical landmark as is its contemporary on Broadway, the Paramount Theater (1931), which was renovated to its original opulent proportions in 1972 and operates profitably today. Possibly, the Fox will be a corner anchor for a huge new $300 million downtown development called the Oakland Galleria, which would occupy the triangular area bounded by San Pablo Avenue, Telegraph, and Twenty-first Street. The zone is now virtually unused with many abandoned storefronts across from City Hall. Although other downtown areas have been successfully revitalized, there are doubts now as to whether the money can be raised to invest in this enormous project.

A classic "flatiron"—the French Gothic-style *Cathedral Building* (1914)—appears on the left as Telegraph merges with Broadway at Fifteenth Street. Another flatiron, the Broadway Building (1907), sits on the right, one block ahead at Fourteenth Street, between Broadway and San Pablo Avenue. These triangular buildings reflect a common turn-of-the-century urban landscape devoted to the streetcar, which had dramatic impact on future urban design. Looking back north from Fourteenth Street, it is easy to visualize three major late-nineteenth-century streetcar lines reaching from the heart of downtown Oakland toward Berkeley: Broadway, San Pablo, and Telegraph. Residual triangular properties at the "gores" are still found in many cities, but rarely are two back-to-back flatirons still in existence. Sadly, the Broadway Building was severely damaged in the 1989 earthquake and is likely to be demolished unless an investor can be found to pay for its repair.

The route now pirouettes past Oakland's City Hall, proceeding up San Pablo to Nineteenth Street. A left on Martin Luther King Jr. Way passes a contemporary urban showplace at Thirteenth Street, *Preservation Park*. Opened in the summer of 1991, this block of sixteen Victorian-era homes has been renovated and put on display as a "collection" of presentable—even fashionable—office buildings and entertainment halls available for wedding receptions and the like. Surprisingly, most of this "neighborhood" is a complete fabrication, including the lamp posts, mid-street fountain (made in 1880s Paris), and the eleven buildings that were moved here when

The flatiron Cathedral Building at Broadway and Telegraph in Oak-
land—an intersection shaped by early twentieth-century streetcar lines.
Photograph by Robert A. Rundstrom.

I-980 (locally known as the Grove–Shafter Freeway) was built through the city. In reality, this block does not contain meaningful historic landmarks so much as it showcases the idea of urban preservation, while shamelessly promoting the work of designers and builders.

WEST OAKLAND

After crossing over I-980, you enter *West Oakland*, the city's first large residential neighborhood. Many fine, pre-1906 houses remain on Myrtle Street and Sixteenth Street. Early settlers were Greek, Italian, and Portuguese sailors and longshoremen employed at the nearby docks.

Oakland is sometimes called the most integrated city in the United States. Indeed, although a majority of current residents are African American, only 2 percent of these Oaklanders live in exclusively African-American neighborhoods. Thus, although dominant in West Oakland, lower- and middle-class African Americans are also neighbors to Asian, Hispanic, and white residents of the same economic bracket here. Oakland also manifests an uncommon level of community pride in contrast to many inner-city neighborhoods. For example, notice the relatively clean streets and well-kept yards. While there is certainly dilapidated or abandoned housing in this community, most of the homes appear cared for.

Poplar Street marks the beginning of a mixed industrial/residential area (note the rail tracks in the streets). Nabisco was a major employer here until it departed in 1991. The American Steel Company still operates on Eighteenth Street. Turn from Poplar onto Eighteenth and notice the carefully tended Hollywood junipers in the sidewalk, softening the corrugated sheet walls of American Steel's warehouse. Such attention to street-side vegetation is rare in industrial areas.

Cypress Street is noteworthy, first for being renamed Mandela Parkway in 1990, in honor of South Africa's civil rights leader. In 1991, the street temporarily carried both names to reduce confusion. Across Mandela lies I-880 and the infamous Cypress Street Cut-off, which received world-wide television coverage during the 1989 earthquake when this entire section of freeway from Seventh

Street to Thirty-fourth Street "pancaked,'" killing or injuring more than 100 rush-hour motorists. The community is still experiencing repercussions from that tragic event.

The landmark *Sixteenth Street Station*—with its 60-foot archways of Sierra Nevada granite—suddenly appears on the other side of Wood Street. Built in 1912, the station was the western terminus for transcontinental passengers bound for Oakland or San Francisco. In its heyday, more than 50 mainline trains per day rolled through at ground level, while nearly 500 local electric trains stopped every day on an elevated platform. It is still the main Oakland/San Francisco stop on Amtrak's West Coast route, confirming Oakland's longtime status as a key transportation hub (but earthquake damage has pushed service into temporary buildings).

Seventh Street is the central business district for West Oakland. Its role as an important cross-bay transportation artery is marked by the BART tracks overhead, anchored in a median strip bearing the remains of the railroad that formerly took passengers to ferry boats at the end of Seventh Street. The street is also marked by alternating vacant and revitalized small businesses, and community support clinics. For example, the "Disaster Relief Unit of the Western Service Worker's Association" between Willow and Campbell Streets provided relief for West Oaklanders following the 1989 quake. Note the false fronts on the row of abandoned nineteenth-century houses near Chester Street.

Across Mandela Parkway, Seventh Street passes the protruding stump of I-880 before entering an area of low-income housing projects. The most notorious of these, the *Acorn Housing Project,* was declared uninhabitable by the federal government in 1991. Massive deterioration resulted after the 1986 Tax Reform Act— passed during the Reagan administration—removed tax incentives for absentee landlords in these federally subsidized projects. The Acorn Plaza shopping center nearby has also been struggling as local shopkeepers fight against crime to stay in business.

WATERFRONT

The quaint shops of *Bret Harte Boardwalk,* on the right side of Fifth Street just after Jefferson Street, are located in a row of old

Heinold's First and Last Chance saloon—an old Jack London hangout. Jack London Waterfront, Oakland. Photograph by Paul F. Starrs.

Victorian homes renovated in 1962, the first such renovation in the United States. Another renovation project, the old Western Pacific railroad depot on Third Street, was donated to the Oakland Chinese Community Council for use as an Asian adult day-care center in 1991.

A right turn toward the foot of Broadway brings you into *Jack London Waterfront,* an area long popular as an entertainment, restaurant, hotel, and boating center on a portion of Oakland's lengthy waterfront. Long known as Jack London "Square," the area's new name reflects the city's intention to capitalize on its proximity to the bay. Like its counterpart in Monterey's Cannery Row, Jack London Waterfront also commemorates a famous author by applying his name and book titles to tourist-oriented businesses, and by the production of select local myths. For example, the log cabin near Heinold's First and Last Chance Saloon is reputed to be London's Yukon home, but is actually a fake.

Heinold's is real, however, and visitors are encouraged to step inside for a cold beer. This was a haunt of London's when he periodically returned to live in Oakland. In keeping with that heritage there is no electricity, no plumbing, a very slanted floor, and a bartender usually willing to spin wild yarns about London and colorful aspects of Oakland's history.

The next part of the route winds through Oakland's *Produce District,* which is scheduled for abandonment after 1991. The four-block area bounded by Webster Street, Broadway, First Street, and Third Street was under single ownership when the primarily Italian and Asian merchants were evicted. It is likely to be incorporated soon as part of the nearby entertainment district.

The one-block area west of Broadway between Eighth and Ninth streets is in a remarkable state of transition. A number of buildings from the 1870s and 1880s have been successfully renovated and leased for upscale uses; the Fana Ethiopian Restaurant occupies one of them at 456 Eighth Street. Across the street lie the ruins of an old burlesque hall, the Moulin Rouge, whose Red Mill sign reclines in the gutted interior while developers decide its future.

Around the corner on Washington Street, Ratto's Delicatessen is a longtime landmark of downtown Oakland. Weekly "opera nights" feature singing table attendants serving thrilled customers at tables covered with red-checkered tablecloths. Finally, Victorian Row on Ninth Street preserves the much-admired Nicholl Block, built in 1876. In 1991, the buildings housed retail antique stores below and offices above.

ASIATOWN

Across Broadway lies the commercial heart of Oakland's bustling, sprawling Asian community. In fact, there are now three separate "Asiatowns" in Oakland (the others are on East Fourteenth Avenue). This one is frenetic with shoppers of all ages and ethnic backgrounds perusing merchandise, including fresh produce stacked on the sidewalks every day (Saturday morning is especially busy). Increasingly, this is the shopping destination for Chinese San Franciscans as well.

The Pacific Renaissance Plaza occupying the block north of Ninth Street between Franklin and Webster streets is a unique multi-use development with an Asian emphasis. Office, retail, and (valuable) parking space are integrated with 250 condominiums and rental units, a library, an Asian cultural center, and a Cantonese restaurant—financed primarily from Hong Kong with additional support from the city of Oakland. The unique project, scheduled for completion in 1992, ensures a more than symbolic link between recent Asian immigration and the revitalization of downtown Oakland. Korean developers are already looking to establish a similar "Koreatown" in Oakland for the 13,000 Koreans who have immigrated to the Bay Area in the last decade.

LAKE MERRITT

Turn down Oak Street, pass the uniquely designed *Oakland Museum* and the *Alameda County Courthouse,* and emerge onto Lakeside Drive. On the right, Lake Merritt is actually a 155-acre saltwater tidal basin that was transformed into the country's first urban wildlife refuge more than a century ago. Recognizing the lake's importance as a continued environmental "benchmark" for the metropolitan area around it, the California State Coastal Conservancy recently proposed a plan for enhancing the lake and its surrounding park.

Development around the lake has always been low-key despite the presence of the Kaiser Center on its western arm. Thus, the fifteen-story *Bellevue-Staten Condominum* tower at 492 Staten Avenue has been an important landmark since construction in 1929. Its Spanish baroque–art deco exterior makes it the most elaborate residential tower in the East Bay. Notice also, the humorous "hinged" exterior on the house at 461 Bellevue Avenue. Unfortunately, I-580 was built between Lake Merritt and Oakland's earliest existing movie palace—the *Grand Lake Theater* (1923)—thereby obstructing an impressive, all-at-once view of this landmark. It is well worth asking the ticket vendor for a quick glimpse of the recently refurbished interior. Be sure to see the Egyptian room as well as the main theater. Next door, the renowned Kwik-

Way, one of the earliest drive-in fast-food restaurants in the Bay Area, provides fine milkshakes and french fries.

PIEDMONT

In the *East Bay*, as in San Francisco and most urban areas of the United States, residential prestige and median household income rise with elevation. Pass the theater, staying on Grand Avenue, and pass through a small shopping district serving one such middle- to upper-class hill-oriented neighborhood. You enter the city of *Piedmont* after the intersection of Grand and Wildwood Avenue. Wholly surrounded by Oakland, Piedmont has long been a bastion of the East Bay upper class. Within easy walking distance directly west of here is the Rose Garden, a small, beautiful park with lanes graced by hundreds of different kinds of award-winning roses.

A right turn from Grand onto Moraga Avenue puts you between Piedmont on the right and *Mountain View Cemetery,* another Frederick Law Olmsted creation, on the left. The high wall shields the view, but the Piedmont Avenue entrance back down the hill opens into this monumental necro-landscape. Huge crypts mark the final destinations of the living members of many of the Bay Area's historic and wealthy families. Moraga Avenue winds along the northern margin of Piedmont. Visitors interested in magnificent residential architecture should meander through the hills on some of Piedmont's sinuous streets. Glen Alpine Road offers some especially breathtaking mansions and vistas.

Moraga rises eastward up a shoulder of the Oakland Hills until, at Estates Drive, it dips down toward the Warren Freeway (Highway 13). You will see evidence of the terrible fire that swept through the Oakland-Berkeley hills on October 20, 1991 and destroyed 3,600 homes.

Veer right onto the southbound entrance and proceed onto the highway. The road travels in a small, narrow valley below which lies the Hayward Fault, a branch of the San Andreas Fault running up the east side of the bay from the Hollister area through the campus of California State University at Hayward, through these hills, north underneath Memorial Stadium on the University of California campus, and beyond.

ANTHONY CHABOT AND REDWOOD REGIONAL PARKS

A pleasant side trip can be made by taking the preceding exit to Redwood Road heading north. Redwood Road leaves Castro Valley, winding its way back over the hills from the east side, eventually taking you back into Oakland via Anthony Chabot Regional Park and Redwood Regional Park. Near the northeast is the Las Trampas Regional Wilderness, which is best reached from the town of San Ramon. A large share of municipal water supplies for the smaller Bay Area cities are stored in reservoirs in these parks. Over a million acres of scenic regional parks form a "greenbelt" through the nine counties surrounding the bay, with numerous peaceful hiking, bicycling, and equestrian trails.

Montclair, a large, residential area for Oakland's wealthy, extends upward into the hills to the left. Many homes are cantilevered from the sides of hills in pursuit of panoramic views. Many of these slopes are unstable, however, and several landslides have occurred in recent years.

An unusual juxtaposition in the religious landscape of the Bay Area looms ahead on the right as we head south on Highway 13. The massive concrete monolith of the Mormon Temple rises on the hill where Joaquin Miller Road and Lincoln Avenue cross the highway. Immediately next to it on the downhill side is the copper-roofed Greek Orthodox Church. Tours are available of both of these Bay Area landmarks, built just three decades ago.

Highway 13 ends and merges onto I-580 near *Mills College,* a private school for women. Continue south on I-580 through East Oakland and San Leandro. Traditionally, the central part of Alameda County south of Oakland has had a large Hispanic minority population. Though still the majority, the white population is in decline, and Asian and African-American ethnic groups have more than doubled here since 1980.

I-580 veers east just beyond the San Leandro city limits. Move into the left-hand lane to prepare for this junction. The freeway penetrates the first ridge of the East Bay Hills and emerges into Castro Valley, longtime upper-middle–class, predominantly white commuter suburb of the East Bay. Usually temperatures immediately begin to increase, and conditions become more arid as travelers leave the immediate bayside. To follow the route, take the Center Street exit off I-580, and proceed toward Livermore.

Livermore

Livermore Avenue crosses I-580 and proceeds toward downtown *Livermore,* becoming North Livermore Avenue. Formerly a small agricultural community from which special orchard crops were shipped, Livermore is now a burgeoning mid-distance suburb in eastern Alameda County. The downtown has been transformed and reflects the influx of young families in the past two decades.

Although a welcome sign on the west end of town announces, "City of Livermore, Wine Country Since 1849," grape production here was never very significant until the last thirty years. Now, several large winemakers are located here in wineries close to the metropolitan Bay Area. Wineries offering free tasting can be reached by turning right from North Livermore onto Fourth Street, driving three blocks, and then turning left (south) onto Arroyo Road and proceeding approximately 2.5 miles.

Turn left from North Livermore onto East Street. East Street provides a good vantage of the juxtaposition of old and new land uses in Livermore: 1960s lath and plaster, ranch style housing on both sides of East Street alternates with remnants of old orchards. Individual fruit or nut trees still visible in the backyards give evidence that many of these suburban residences were aligned and superimposed on the preexisting agricultural land. More wineries are located less than 2 miles away, on Tesla Road, a southern

East Bay to Mariposa

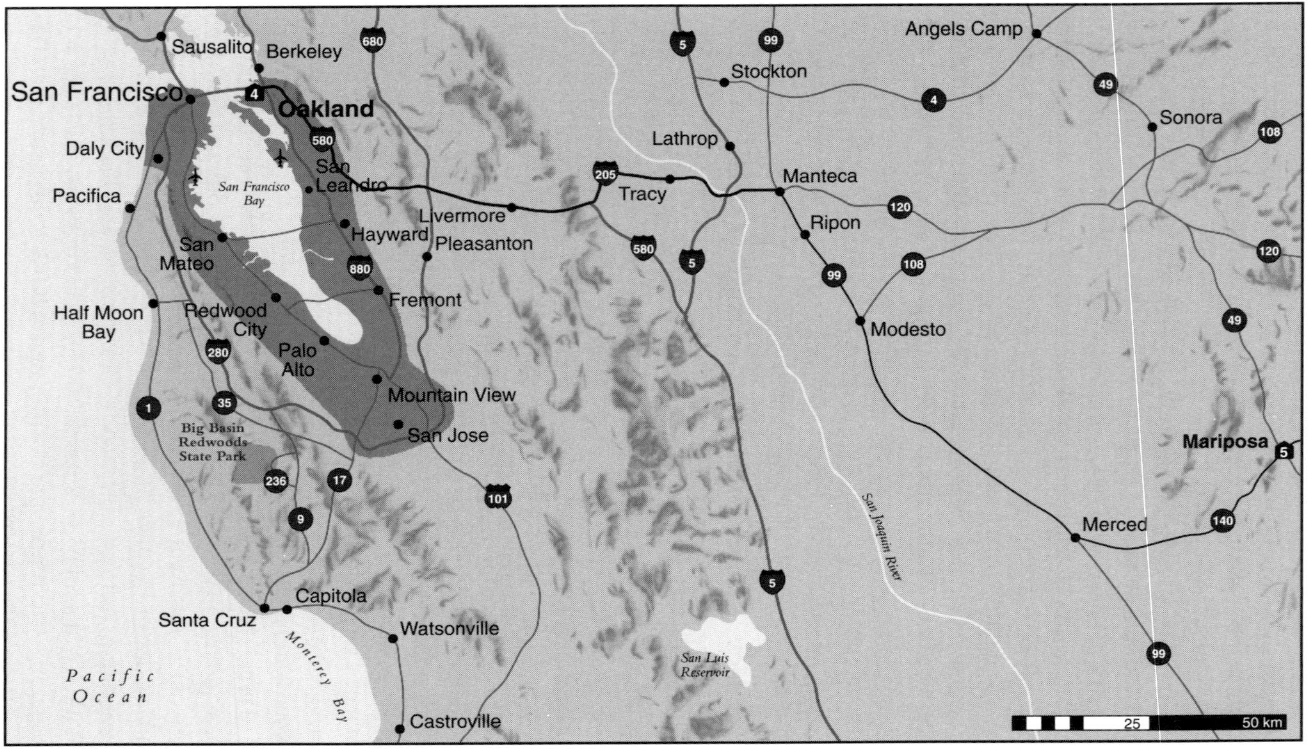

parallel to East Street. Coal mining in the early twentieth century was locally important in the hills to the far right.

Two noteworthy aspects of the technological landscape in the Livermore area are the famous (or infamous) *Lawrence Livermore National Laboratory,* once known as the "Rad Lab," on the left after crossing South Vasco Road, and the Hetch Hetchy Aqueduct. The massive laboratory facility is operated by the University of California at Berkeley for the federal Department of Energy. The word "radiation" was dropped from the laboratory's name and from the sign on the left side of East Street during the social turbulence of the early 1970s. The Hetch Hetchy Aqueduct, San Francisco's municipal water supply, passes through the fault-ridden hills to the south, approximately 10 miles from this intersection.

The Lawrence Livermore Lab's Visitors' Center is on the left after turning left at the stop sign on Greenville Road (the Molzaha Ranch is directly ahead at the intersection). Notice the double row of eucalyptus and pine trees serving as both camouflage and windbreak on the left. An operating ranch across the street from a top-secret weapons research facility is perhaps typical of the odd juxtapositions on the California landscape.

Wind Power

From here, some of the 5,000 wind turbines of a gigantic "wind farm" are visible ahead on the right, atop the hills at Altamont Pass. The route now leads in that direction, passing under the Union Pacific railroad tracks, and again under I-580, before turning right onto Old Altamont Pass Road.

The Old Altamont Pass Road ascends through a narrow extremely windy canyon bounded by working ranches, sere grasslands (some parcels on the left are notably overgrazed and eroded), wind turbines, and railroad tracks on both sides. Notice the cattle trails on the hillsides. Are freeways important to Californians? Observe the expense involved in supporting a short, elevated section of I-580 up the steep slope to the right.

Windfarm at Altamont Pass. Photograph by Paul F. Starrs.

Across Dyer Road is the abandoned old site of the town of *Altamont*. The ruins of the Summit Garage are about all that is left. In 1970, the last multi-day outdoor rock music festival was held near here, ending with the on-stage stabbing of one reveler by members of the Oakland-based Hell's Angels motorcycle club as the Rolling Stones continued to play.

The summit here is low, less than 1,300 feet, but the whole region is notoriously and perennially windy. Approximately 60 percent of California's wind-power generating-capacity is located here. The generators are operated by several private companies, including San Francisco's U.S. Windpower Inc., the largest wind company in the state, and Fayette Manufacturing Co. in Tracy, which operates about 1,650 of the turbines here. Flowind Corporation in Pleasanton (west of Livermore) builds and repairs many of the generators.

Drive onto the gravel shoulder, park the car, shut off the motor, and listen! This is the world's largest concentration of wind turbines. Ten percent of California's electricity demand was supplied by these turbines between 1980 and 1985. But the fledgling industry sagged when the Reagan administration agreed with the outcry of oil companies and eliminated the federal tax credits that had supported wind farms. This and other wind farms limp along today on what little state support they can gather. However, recent technical improvements have made these generators cost-competitive with oil or coal-fired power plants.

Altamont Pass to San Joaquin Delta

Turn left onto Grant Line Road at the stop sign at the bottom of Altamont Pass Road. Within 1.5 miles, the road crosses over two other notable features in California's techno-landscape—the *California Aqueduct* and the *Delta-Mendota Canal.* Water from these two southward-flowing water-transport systems feeds much of the San Joaquin Valley's famous agricultural productivity (for instance, Fresno County agriculture is the highest in cash value of any U.S. county), on which many Americans depend for fresh fruit, vegetables, nuts, and specialized cash crops.

The California Aqueduct moves water pumped from the large Clifton Court Forebay just 7 miles north of here. The Delta-Mendota Canal returns north-flowing San Joaquin River water back into the valley. Both systems are part of an elaborate network of sloughs, distributary streams, canals, and natural and artificial levees that make up the Delta—the large Sacramento–San Joaquin Delta. Peat soils rich in organic matter, particles of which are easily wind-borne, and sub-sea-level islands are also common here.

Before the introduction of large-scale, mechanized agriculture, the west slope of the Sierra Nevada to the east, and the entire Central Valley drained into San Francisco Bay via the Delta. Now, various water interests in the southern San Joaquin Valley and the Los Angles Basin have succeeded in diverting much of this fresh water southward via these two aqueducts. Without freshwater flushing in the Delta, saltwater incursion from the bay is an increasing problem that, ironically, threatens

CALIFORNIA AGRICULTURE

California is the leading agricultural state in the nation. In 1990, farmers in California received 12 percent of all national farm revenues, or $17 billion. Eight of the top ten agricultural counties in the United States are in California. Fresno County, the number-one agricultural producer in 1990, earned over $3.5 billion in farm revenues, principally from cotton and grapes. California is among the nation's leading exporters of farm products, sending over $5 billion worth of farm products to foreign countries.

Agriculture ranks with aerospace and electronic industries as cornerstones of the state's economy. Yet, unlike its industrial counterparts, contemporary agriculture in California is difficult to assess and interpret because of its diversity, its specialized nature, and its regional variation. Two factors underlie and contribute to the development of California's contemporary agricultural patterns: the nature of the physical environment and the evolution of the farming system.

California's physical environment presents a unique combination of soils, topography, and climate that has created a wide range of environmental conditions within a relatively small area. These localized environments have allowed farmers to take advantage of specific environmental conditions and grow high-value specialty crops. Thus, in many instances crop patterns are influenced more by the dominance of specific physical factors than by economic market conditions. Environmental factors determine what may be grown and, in some measure, to what advantage.

While California's physical environment gives many advantages to the farmer, it also has one distinct drawback, a regional imbalance in rainfall. Most of the state's rain falls during the winter, in northern California and along the Sierra

Raisins drying in San Joaquin Valley. Photograph by Paul F. Starrs.

Nevada. To compensate for inadequate rainfall and to take advantage of local environmental conditions, California farmers have turned to irrigation on a grand scale. They have moved water from regions of high rainfall (where few people live and where the climate is such that few crops can be grown profitably) to regions of low rainfall (where much of the population resides and where the conditions are such that the agricultural potential is far greater). California's environment may dominate agriculture, but it is irrigation that makes agriculture profitable.

California's urban population and industrial economy have grown tremendously during the past twenty years, which has placed considerable pressure on land and water resources. In response to increased demands and fewer resources, farmers have introduced higher-yielding crops, improved farming methods, refined marketing techniques, developed new technology, and in some cases changed location to take advantage of specific environmental conditions that would improve output. The typical California farm also underwent changes in organization, management, and ownership during this period. Many increased in size, focused on a single crop, became corporatized, and were managed by nonfarmers whose interest in agriculture was limited to the costs and market price. Today, farming in California is a business—a huge business—and consequently farms are operated like any other profit-making enterprise.

agricultural productivity along with some commercial sport fish species in the Delta region.

Continue east on Grant Line Road into the valley, crossing the San Joaquin County line and passing through the northern part of *Tracy,* another small, agricultural community that has become a bedroom community of the Bay Area. Daily commutes to the Bay Area are commonly two hours in each direction from here, yet

Sheep grazing, with herder's wagon, in San Joaquin Valley. Photograph by Paul F. Starrs.

some drive from Manteca, and even Modesto, which are still farther east. Stay on Grant Line Road through Banta.

Turn left from Grant Line Road onto Interstate 5 near the Deuel Vocational Institute, a working prison farm. The freeway is elevated above characteristic features of the Delta. Nearby are Tom Paine Slough, Paradise Cut, and Stewart's Tract, an average-sized agricultural island. As the freeway crosses the San Joaquin River, note the three generations of bridges on the left. After passing into the *Lathrop* city limits, exit I-5 at the Louise Avenue off-ramp, and turn right onto Louise.

Three large companies indicative of the agricultural industry appear immediately on the right before crossing McKinley Avenue: Libby-Owens Ford Glass Company, a Best fertilizer plant, and the Simplot Company. Numerous local wells have been drilled to the water table in this area. The first of what will become an abundance of almond trees appear on the left at the intersection of Louise and Airport Road.

Enter the city of *Manteca* (population 30,000) after crossing the Southern Pacific railroad tracks. Formerly dominated by an odorous beef slaughterhouse, Manteca is increasingly the site of affordable family housing for long-distance Bay Area commuters. The elevation here is only 38 feet above sea level. Liquidambar (Sweetgum) trees have been planted in the sidewalk strip here.

Turn right onto North Main Street (Manteca Road) at the stop light, drive through Manteca's small "auto row" (note the emphasis on pick-up trucks), and then turn left onto East Yosemite Avenue (U.S. Highway 120). After crossing Powers Avenue, note the almond trees and Spreckels sugar refinery on the right (recall the company town of Spreckels near Salinas in the Monterey Bay region).

On the east side of Manteca, travelers begin to enter into the heart of one of many important agricultural subregions in California's Central Valley. Between here and Modesto, approximately 15 miles away, almond and walnut groves and irrigation canals— oriented along an east–west axis—dominate the agricultural landscape. Turn right onto Jack Tone Road and drive toward the small town of Ripon just 4 miles away. Irrigation canals are now oriented perpendicular to the path of travel, and must be protected underneath the road pavement. Roadside stands sell assorted fruit and nuts fresh from the fields (in season).

After crossing over U.S. Highway 99, turn left onto West Ripon Road, which then becomes West Main Street in *Ripon* (population 3,500). Ripon is one of the few remaining examples of what a small agricultural town in this part of Central Valley would have looked like thirty years ago. Notice that suburban development has yet to overtake Ripon as it has Tracy, Manteca, and Modesto. In part, this is explained by its being equidistant between Manteca and Modesto, where development has been concentrated. Ripon still rests on its agricultural roots, with the Ripon Milling Company and the newer Nescafé plant employing significant numbers of local residents. However, it is easy to see that Ripon is being gradually squeezed by suburban encroachment from the two nearby

THE SAN JOAQUIN VALLEY: CALIFORNIA'S RURAL MELTING POT

The decade of the 1980s profoundly transformed California's ethnic composition, and no region of the state underwent a greater change than the great San Joaquin Valley. The San Joaquin Valley, the landing point for northern European, Hispanic, Armenian, Mediterranean, and other immigrant groups throughout the majority of the twentieth century, suddenly added Asians to the mix during the 1980s.

Asian refugees began to flood the valley towns after spending unhappy winters in Minnesota and Wisconsin. They came not only for the warmth, but to farm the valley's mythical soil, and to be reunited with family and friends. Today, schools in Porterville, Merced, Visalia, and Bakersfield must contend with students who speak Hmong or Lao, as they learn the basics of American history alongside the sons and daughters of Armenian grape growers, Basque sheepherders, and Mexican fieldhands.

During the 1980s, Asians were the fastest growing ethnic group in many of the valley's towns. Merced, Visalia, Modesto, and Fresno each had well over a 500 percent increase in Asians. As Asians eventually came to outnumber blacks in California, a new look was slowly superimposed on towns across the state's farming heartland. By 1990, the census had identified 175,000 Asians in the San Joaquin Valley, most of whom were born into rural and even tribal Asian cultures. For example, the Hmong, who fought Communists in Laos as mercenaries for the CIA during the 1960s and 1970s, are now a major component of the new Asian population; they are followed by a smattering of Mien, a hill tribe from Laos; the lowland Lao; and Cambodian Buddhists.

While the Hispanic populace remains the largest ethnic minority in the San Joaquin Valley and accounts for almost 32 percent of the valley's total population, Asians rank a distant second with more than 9 percent. As we head toward the middle of the 1990s, much of our attention will naturally focus on urban ethnicity and the attendant problems of the expansive California cities, but the next observable melting pot may well be the state's rural heartland.

cities, and it will likely be absorbed as orchards continue to be converted for housing. Turn right onto Stockton Avenue at the stop sign, then make a quick left onto Second Street past the Ripon Milling Company and Nescafé properties, and veer right onto U.S. 99 south, and into the heart of the renowned *San Joaquin Valley*.

Highway 99 takes you south through Modesto, Ceres (the classical Greek goddess of grain and harvests, a good name for a town in this fertile farmland), Turlock, Atwater, and Merced. In Merced, you will turn east onto Highway 140 to Mariposa, where you enter the Gold Country in the foothills of the Sierra Nevada range.

Mariposa (named for the butterflies that once frequented the area), once the virtual heart of the Sierra gold rush towns, continues to prosper as a service center for the surrounding region, as well as a major fueling point for tourists on their way to Yosemite Park. The widely acclaimed county courthouse is deemed to be the oldest continuously used courthouse in the state. The nearby Diamond W Ranch and Pack Station affords the tourist the opportunity of experiencing veritable Old West gold-mining conditions, and the Mariposa Mine is notable as the first quartz steam mill in California. The state runs a mining and mineral museum 2 miles south of town and in the county fairgrounds.

Highway 49, which links the gold mining towns, goes north out of Mariposa. We will join it later, after following the steady flow of traffic east on Highway 140 to El Portal, an entrance to Yosemite National Park. (An alternate route into Yosemite Valley takes Highway 49 south to Oakhurst, and then Highway 41 north to the Wawona entrance of the park.)

△ *Day Seven*

YOSEMITE AND TAHOE
TO MONO LAKE

Take Highway 140 into *Yosemite National Park* (hold on to your receipt for your entrance fee) and make the loop around Yosemite Valley, and then out again by Big Oak Flat Road (Highway 120) west to the Big Oak Flat entrance. Even if you do not stop along the way or take the two side-trips we describe below, your entrance fee will seem like a bargain.

You will see the eastward branch of Highway 120, which goes through the high country of Yosemite, come off from Big Oak Flat Road. Excellent guides for all of Yosemite abound, and if you have time and the right weather, you should by all means explore Highway 120 over Tioga Pass and down the east slope of the Sierras, the fastest way over the mountains to Lee Vining. The route we follow instead is admittedly much longer and misses some of Yosemite's grandest sights, but it can be driven all year round (except in the very worst weather) and has its own wonders to behold.

For two-thirds of the year, Tioga Pass is impassable. It can happen any time in the fall: a dusting of snow at lower elevations in the Sierra Nevada foothills that implies serious snowfall higher up in the mountains. The U.S. National Park Service, with jurisdiction over transit across Tioga Pass on Highway 120, regulates travel by an unvarying rule: The pass closes with the first substantial snowfall. That yearly decision has a stern finality; only in rare years with an insignificant spring snowpack will Tioga Pass

Yosemite and Tahoe to Mono Lake

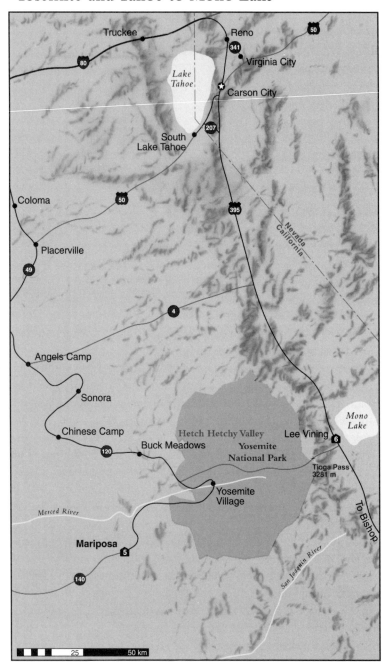

AUTOMOBILES AND YOSEMITE NATIONAL PARK

These soaring cliffs seemed to hold safe from violation the fabled Garden of the West so that it might be there when Californians needed a reminder of what the land offered them, of what it had been preparing for them in the ages before their arrival. In 1864 Americans set the Yosemite aside because it was so unutterably beautiful and because it expressed their own best hopes. (Starr 1981, p. 183)

Undoubtedly, you will drive around Yosemite Valley when you visit Yosemite National Park. If it is summer, you will notice that most of the more than 3 million annual recreational visitors to the park appear to have your same vacation schedule. Observe the abundant tour buses, numerous privately owned stores, hotels, gas stations, and campgrounds, and the twenty-two-bed jail as you contemplate where to look for a parking spot in this mini-metropolis. When you throw your ice-cream-bar wrapper or film container into a "wildlife-proof" garbage can, bear in mind the twenty-five tons of garbage created here on an average day. Is this what Kevin Starr referred to as "the fabled Garden of the West"?

Yosemite can be understood in many ways, but it is worth recalling that ecological issues were of little concern when the U.S. Congress granted 1,500 square miles of the Sierra Nevada to California for a state park in 1864. Lawmakers saw Yosemite as an essentially worthless, unexploitable patch of monumental landscape (development at Hetch Hetchy was still several decades away). What was to become the first national park in the United States was originally considered a pleasuring ground. With unlimited access—a basic tenet of U.S. federal recreation policy—Americans in general, and

Californians in particular, would be inspired to the great achievements necessary for building a world power.

Later defenders of the park's inspiring beauty, John Muir and Frederick Law Olmsted among them, cautioned that an emphasis on overwhelming dimensions could potentially lead to unfettered development and commercialization. Olmsted predicted 1 million visitors by the 1950s, and he feared that without protective laws the park would be overwhelmed and degraded.

Despite these warnings, hardly a tourist brochure was published in the nineteenth and early twentieth centuries that did not promote Yosemite's contrived "natural" attractions. Some, such as the famed "tunnel-tree" at Wawona, required and promoted the automobile as the best way to see Yosemite. The Wawona Tree stood astride Highway 41 at the southern entrance to the park. In 1881, officials hacked a tunnel 8 feet high and 11 feet wide through this 231-foot sequoia. A sign was placed at its base indicating the precise dimensions. For the next eighty-eight years, this creation attracted tourists, including presidents, eager to have themselves photographed in a carriage, stagecoach, car, or small truck as it passed through the opening. Not surprisingly, the Wawona Tree finally thudded to the ground in the winter of 1969.

The public needed to see Yosemite to support the values on display there, and therein lies the origin of many of today's problems. The Yosemite Valley Railroad was brought up to El Portal, 12 miles west of the gorge in May 1907 (the tracks were torn up in 1944). Immediately, preservationists declared an invasion by commerce and tourism. The railroad should have been the least of their concerns for just eleven years later, more than half the visitors arrived by private motorcar, a travel mode that at that time was considered more egalitarian and democratically inspired than the train. In 1918, the proportion was eight to one in favor of the car.

Preservationists were boxed in. Without the automobile, the public would not know of Yosemite's scenic beauty, and there would be little public support for its continued preservation. This remains the Achilles heel of the National Park system today. How else can the American public come to know what it is they have? Thus, the idea of economic development of a recreational resource undergirded public support for the national parks, yet that development inevitably compromised, and in some cases destroyed the values people sought at Yosemite. The debate between preservation and access was on, and without a clear policy, the cars kept coming.

Auto-camping mushroomed in popularity in the 1920s and 1930s. It was considered to be a more authentic, elemental means of experiencing nature compared with the posh elitism then ascribed to railroad travel. By the end of the 1920s, cars were coming so frequently that visitors became more concerned with road improvements and motorists' accommodations than with outdoor or wilderness experience per se. One million visitors arrived by car in 1955 just as Olmsted had predicted, and over 50,000 per day came throughout the 1960s. Today, Yosemite Valley appears designed for automobiles, and air pollution is an enormous problem. But the seeds of Yosemite's dilemmas were planted by the park's founders presiding at its creation.

open again to cars before May 31, Memorial Day. True, Tioga Pass (elevation 9,945 feet) is the southernmost of six routes across the central High Sierra, and the next highway crossing is well south of Bakersfield. But while the Tioga road is a pleasant drive, crossing to the eastern edge of California with only half a day's travel, it's not irreplaceable.

Yosemite National Park. Photography by Martin S. Kenzer.

Anyone who wants or needs to cross the Sierras during the eight months that Tioga Pass remains closed will find alternatives aplenty—the route sketched here cuts through the best of California's remarkable Gold Country, across arguably the most famous pass (and all-weather route) in the United States, and down the east side of the Sierra Nevada, to rejoin Lee Vining at Mono Lake. If the only object is getting there fast, go along Highway 49 until it connects with Interstate 80 in Auburn, cross Donner Pass on I-80 to Reno, and then drive south on Highway 395 from Reno to Mono Lake. Although do-able in one exceedingly long day, doing it in two allows for detours that are well worth taking. These save the journey from being just a demonic mountain driving trial. Reserve moments for photographs, walks, talk in small towns, time to observe helper-engine-assisted trains plodding steadily through Truckee, or taking refreshments in Nevada's oldest bar.

PLUTONS

Plutons form the skeletons of many of the mountain ranges that we encounter on this California expedition. Plutons (named for the Roman god of the netherworld, Pluto) are large, crystalline, intrusive rock masses. The largest plutons are also termed batholiths (from the Greek, meaning deep rock). Most of the plutonic bodies we see exposed in the California mountains are granitic, identifiable by their easily visible light coloration and coarse grains.

Igneous plutons are formed when rising magma displaces surrounding bedrock, forming a sort of molten bubble within the upper crust. Formation of the magma bodies is frequently associated with the subduction and melting of crystal plate materials. Thus they commonly occur near plate boundaries, where the incidence of earthquakes and volcanic activity is also increased.

Because the magma does not reach the surface, the molten rock cools slowly, allowing selective "freezing" of minerals and the growth of large crystals. The large crystal size is readily apparent in the granite exposures in the Sierras. The coarse appearance contrasts sharply with the fine crystal structure of extruded (and thus quickly cooled) basalt, or the glass-like obsidian.

We can also visualize the magma, rising through the surrounding rocks, behaving like cumulus clouds rising through air. Giant bubbles billow from the main mass, creating an irregular, but somewhat smooth, exterior. This form is preserved when the magma hardens. The crevasses in the upper surface of the plutons remain filled with the original, surrounding rock, and have the appearance of roof pendants, as they are termed. The plutonic rocks are exposed when surficial processes remove the bulk of the overlying materials.

We see the resulting granitic exposures at several locations on this trip, in the Santa Lucia Range, for example. However, the most impressive outcrops occur in the Sierra Nevada Range. The spine of these mountains is a batholith composed of numerous plutons whose size and shape control much of the renowned scenery of Yosemite Valley and the western slopes of the Owens Valley. The more than 100 plutons of the Sierra Nevada batholith are believed to originate from the melting of descending edges of the Pacific plate. The oldest of the plutons was intruded approximately 200 million years ago, with later formations occurring as recently as 75 million years BP. Indeed, these rocks are much older than the mountains themselves. The plutons are widely exposed, representing more than 50 percent of the surficial rocks within the Sierra Nevada. From the Glacier Point overlook, in Yosemite National Park, many granitic surfaces reflect the irregularity of the plutonic mass, as evidenced by the occurrence of the spectacular domes.

Park Environs: Tuolumne Grove and Hetch Hetchy

Leave Yosemite Valley by the Big Oak Flat Road, a westward leg of Highway 120. The road's sinuous turns are easily traced on the large-scale Park Service map of Yosemite that visitors receive for paying their fees at the park entrance. Within forty-five minutes of leaving Yosemite Village are two fine side trips. One travels the Old Big Oak Flat Road, a one-way westerly drive descending through the Tuolumne Grove of giant sequoia trees. Ask before leaving the Valley if the Grove road is open (it closes after heavy snowfall). If all is well, then simply make a right turn onto the Tioga Road, before reaching Crane Flat, go less than a mile, and

TEN VIEWS OF A REDWOOD FOREST

Describing California abuses the superlatives in our language. Visitors are accustomed to hearing that the state has the tallest, highest, lowest, oldest, saltiest, youngest, newest, wealthiest, trendiest, weirdest, or most diverse collection of nearly everything imaginable. It is easy to become jaded if you stay in this state very long, and therein lies a danger.

Trees are a good example of the problem. The Earth's longest living (bristlecone pine), largest in girth (sequoia), and tallest (redwood) are found within California's borders. Quoting such uniqueness becomes so common it trivializes meaning. Ronald Reagan demonstrated the problem clearly enough during his 1966 gubernatorial campaign. In response to a proposal for a Redwood National Park, Reagan yawned and declared, "You know, a tree is a tree—how many more do you need to look at?" (in Nash, p. 161). The park was created in 1968 despite the governor's indifference.

A redwood forest is not simply understood. The trees play at least ten different roles in the daily California drama, only one of which refers to their height. These roles correspond to human viewpoints or perspectives that are often in bitter opposition. Brief descriptions of these perspectives form a sort of social hall of mirrors. Everywhere you turn, there appears a conflicting, but socially valid image of a fairly simple biological specimen.

First, there is of course the *monumentalist* perspective in which the sheer visual impact of a redwood's vertical dimension—many are taller than 300 feet—and often its girth, are the only significant aspects of the tree. Some see redwoods as a freakish curiosity of nature, suited for commercial exploitation as drive-through tunnel-trees of which there are at least three still remaining on private land (do not confuse these with the Wawona Tree, which was a *Sequoia*). Generations of professional and amateur photographers are the "caretakers" of this perspective. Many such caretakers can be seen in action beside—and inside—a redwood trunk on display at the Ayala Vista Point at the north end of the Golden Gate Bridge, gateway to the Redwood Highway (the northern portion of U.S. 101).

The second perspective is that of the *feller*, the person in the logging industry who brings the trunk to the ground so that the tree and everyone in the vicinity are left intact. This is no easy achievement and often the feller has great respect and admiration for redwoods, even while the trees are viewed simply as a source of seasonal income.

Third is the *transcendentalist* who seeks peace and tranquility alone in the forest. Like Thoreau and Emerson, the goal of today's transcendentalist is to release the mind from the pressures of everyday life, to transcend mere existence so that a clearer vision of reality and human nature is possible. Quiet and contemplative, the forest provides stimulus for deep, philosophical thought.

Fourth is the *timekeeper.* Scientist and amateur alike find meaning in the redwood by counting its annual growth-rings to find its age. The tree's past is often linked with events in human history by a crude time chart displayed next to a slice of the trunk so, for example, the viewer can find the ring corresponding to the year Alexander the Great was born. You can see a large "time slice" at Muir Woods National Monument in Marin County, just north of San Francisco.

Redwoods symbolize the power and strength of the nation in the fifth view of the *national supremacist.* In the nineteenth century, it was common to tout the trees to Europeans as a living antiquity without equal in Europe, a testimony to the greatness of America that is occasionally still heard today. This boastful perspective "proves" Californians have roots in the ancient past, which still grow, unlike the walls of ancient cathedrals.

The *isolationist* holds a sixth perspective. The forest represents an effective shield from the outside world, a place to get lost, to shed personal emotional burdens (see John Lithgow's fine portrayal of a forest-dwelling Vietnam veteran in the 1988 movie, *Distant Thunder*). Redwoods also provide refuge for those operating outside the law. For example, marijuana is the primary cash crop in Humboldt and Mendocino counties (two thirds of "The Emerald Triangle") due to the productive—and well armed—plantations secreted away in isolated mountain valleys.

The seventh perspective is that of the *ecologist,* the dispassionate scientist studying the redwood's niche in the web of life. The ecologist sees the tree as but one element of a highly integrated ecosystem in which fire, periodic flooding, and fog are equally crucial. The dimensions and age of the trees are simple manifestations of the system's long-term stability in the face of periodic change.

Eighth, the *preservationist* sees the redwood as a symbol of the power and fecundity of nature and the comparative insignificance of human existence. The forest rightfully dominates as a kind of temple or cathedral requiring extreme reverence. Many backpackers and hikers, and some writers and publishers hold this view. John Muir is a folk hero for preservationists, although Muir was an astute ecologist as well.

A ninth view is that of the *forester* who sees redwood trees as contributing a number of board-feet at maturity. The tensile and torsion strength of the wood, and its legendary resistance to fire, pests, and rot are measurable properties of great interest.

Finally, there is the perspective of the *financier,* including the owners and stockholders of lumber companies, and junk-bond brokers, all of whom see redwoods as a liquefiable financial resource. For example, in 1985, when Maxxam Inc. acquired the venerable Pacific Lumber Company in Scotia (Humbolt County), the controlling financier, Charles E. Hurwitz, ordered clear-cutting of the largest stand of redwoods in private hands in order to pay off debts the company took on to finance the purchase.

Hetch Hetchy Reservoir. Photograph by Martin S. Kenzer.

turn left into the Tuolumne Grove parking lot. Buses and vans, forbidden to travel the road, always discharge some fanatical visitors who walk the several steep downhill miles to the Grove; car traffic is sparse. The drive, though, is well worth a bit of careful driving and a few white-knuckle hairpin turns. The giant sequoia (*Sequoia gigantea*) is one of the two varieties of California redwoods; these are the montane variant, while the coastal redwood (*S. sempervirens*) rarely appears more than a dozen miles inland. Any car of reasonable size can negotiate a tunnel carved through one of the Tuolumne Grove trees. After the hubbub of people in Yosemite Valley, it's restful to squeeze off the road into a parking space and wander in incomparable silence, surrounded by huge trees. The winding road discharges near the Big Oak Flat entrance station, where your park fee slip must be presented before leaving. Just beyond that is an attractive second detour from the main route.

THE HETCH HETCHY DEBATE

The Gold Rush and ensuing influx of immigrants had left San Francisco with a population of 350,000 by 1900, and without a reliable water supply to support it. In 1901, the city engineer issued a proposal to dam the Tuolumne River at the head of the dramatically scenic Hetch Hetchy Valley 150 miles distant: The water was "free" since it was already in the public domain. The secretary of the interior in Teddy Roosevelt's first administration rejected the plan, arguing that there were adequate sources closer to the city.

The 1906 earthquake occurred with perfect timing. San Francisco was razed by a massive fire following rupture of the local water supply, which was stored in sag ponds over the San Andreas Fault zone. In 1908, during the second Roosevelt administration, the plan was resubmitted and approved by Secretary Garfield, who wrote: "Domestic use is the highest use to which water and available storage basins . . . can be put" (in Nash, p. 161). His view reinforced that of the plan's proponents who argued that the needs of a city supersede those of a few "wilderness-lovers," and that a lake would enhance, not destroy, the scenic beauty of the valley.

Meanwhile, John Muir and others joined battle and succeeded in stalling the project by using powerful religious imagery in a kind of holy war. But, ultimately, the forces for development were victorious, and construction began under the Raker Act of 1913. The reservoir began filling in 1923, setting a crucial precedent because it was not meant to water an existing urban population, but a future one. Today, Hetch Hetchy Reservoir supplies most of the water, and much revenue from the sale of hydroelectricity, to San Francisco and the South Bay Area.

For the first time in U.S. history, a national controversy over the uses of nature had aroused passions at the highest levels of government. The legislature was especially upset because virtually every western congressman who voted in favor of the city in 1913 seemed tormented, prefaced his vote by proclaiming a love of wilderness and a reluctance to destroy it. All agreed it was a debate between two "goods," not between "good and evil."

The controversy is not over, however. On 29 July 1987, Interior Secretary Donald Hodel suggested a federal study be conducted to determine the feasibility of destroying the dam and restoring the drowned glacial valley behind it so Hetch Hetchy could serve as a spill-over site for the 3 million visitors who annually overrun Yosemite Valley.

The Hodel proposal reignited the debate over Hetch Hetchy. A study by the U.S. Bureau of Reclamation suggested that the reservoir could be made superfluous by increasing storage in the three other large reservoirs built on the Tuolumne River after Hetch Hetchy. The city of San Francisco was opposed, however, because it relies on selling the dam's "free" power to offset budget deficits. In June 1988, the U.S. Congress denied Hodel funding for the study, and the matter has been temporarily laid to rest. But the symbol of Hetch Hetchy will live on in the continuing debates over the best and fairest use of flowing water and scenic landscapes.

A right turn past the Big Oak Flat park entrance moves you onto Evergreen Road. Congenial enough on its own merits, the road's importance is mainly as a conduit to *Camp Mather* (maintained as a summer camp by the City of San Francisco). Just after the camp is a junction with the Hetch Hetchy Road. Once a mule trail, then a standard-gauge railroad line, and only lately a perfectly adequate two-lane paved road, the highway dead-ends at *Hetch Hetchy Reservoir.* Yosemite Valley lies on the Merced River, and has universal acclaim as the best-known example of a glacially shaped valley in the world. But a near twin to Yosemite is Hetch Hetchy

Valley, only a short distance north of Yosemite on the Tuolumne River at the head of the Poopenaut Valley.

The valley's first known non-native visitor was Joseph Screech, who came to the valley in 1850, intrigued by the possibility of grazing livestock on the silky green meadows kept free of trees by Miwok Indians' burning (the valley is named for a grass cultivated for its seeds). An alternative use soon presented itself. John Muir always claimed Hetch Hetchy Valley surpassed Yosemite's beauty. Since the city of San Francisco completed O'Shaughnessy Dam at the mouth of Hetch Hetchy Valley in 1938, any debate comparing the valleys is academic; the floor of Hetch Hetchy Valley today lies under 308 feet of water (more or less, depending on water reserves). Stories about Hetch Hetchy, pioneer conservationist John Muir, the political birth of the Sierra Club, and the astonishing inefficiency of San Francisco in building the dam abound. Probably construction of Hetch Hetchy did break John Muir's heart and kill him. Certainly San Francisco's reach to the Sierra Nevada for a secure water supply, including building a dam and reservoir within a national park, desecrates a spectacular wild valley, and constitutes a bit of piracy no less brazen than the city of Los Angeles's building of the Owens Valley aqueduct. What California cities need they take.

Fortunately, while Hetch Hetchy Valley is drowned, its tiny reservoir area (3 square miles of water surface—less than 2,000 acres), tall faces of sheer granite, and Hetch Hetchy Falls pouring into the reservoir are at the least a feast for the senses. This is both grand scenery and a slice of California history (ask for information sheets at the Hetch Hetchy entrance station). And dams, up close, acquire a substance and majesty that only the most hardened dam-hater can surmount. O'Shaughnessy lacks Boulder Dam's art deco elegance or Glen Canyon Dam's concrete-to-sandstone bravado. It does all right, nonetheless. A walk across the top of the dam face, through the granite tunnel on the far side, and a pause to watch the downstream rooster-tail of water crashing from the penstocks into the opposite side of the riverbed are obligatory. In contrast to Yosemite Valley, everyone respects quiet. Return as you came; there are no better short-cuts back. Follow Highway 120, the Big Oak Flat Road, to the north and west, leaving Yosemite National Park behind.

Coulterville or the Old Priest Grade to Chinese Camp

From the Yosemite Park entrance the Big Oak Flat Road drops and climbs the rough several branches of the Tuolumne River. This roller-coaster ride is typical of all Gold Country roads, and the higher a route in the mountains, the more abrupt each descent and reemergence. Ahead, within 35 road miles, lies Highway 49, dubbed the Gold Rush Highway by foothill realtors and the Transportation Department officials in charge of such matters. For once, the name isn't bad; Route 49 passes near or through every substantial nineteenth-century gold-mining town in the Sierra Nevada foothills. Sun-drenched in summer, chilly in winter, this region has long been a handsome, even quaint, part of California that only lately has raised its head against the weight of history. While some of these places feast on tourists and have an unmistakable ticky-tacky flavor, a few simple side-trips reach an easygoing side of California that echoes silent dignity.

The Sierra Nevada foothill counties are enmeshed in a love–hate relationship with urbanization. Second homesites (many not built upon) sprouting like mushrooms in the 1950s and 1960s began the growth; city folks fleeing Sacramento, Fresno, and even the San Francisco Bay Area are finishing the job, moving to the rolling oak-, pine-, and chaparral-dotted hills as if back to the good earth. Not until 1990 could every foothill county finally claim to have more residents than during the Gold Rush era, 140 years before. A consequence of all this is patchy development. For mile upon mile, vegetation, too, looks like a vast and complicated mosaic: the mixed species that make up California's chaparral, with pines and oaks rising above the highly flammable shrub understory. As the foothill country is reoccupied by modern urban refugees escaping the cities in search of the good life, new burdens are placed on the hinterland. Not the least of problems is an escalating risk of catastrophic fires—the risk builds not because fires are any more likely, but because the steadily rising population is more susceptible, living in subdivision homes along roads that often have only one exit—through the gated community entrance. For every modernizing boom town, another turns its back, attempting a return to hibernation; the spectrum is broad along Highway 49's transect.

The Big Oak Flat Road, traversed by horse-drawn coaches bringing sightseers to Yosemite a century ago, joins Highway 49 at *Moccasin*. The last few miles before the junction drop eerily from 2,800 to 936 feet at the Moccasin Powerhouse, where water diverted from Hetch Hetchy plummets through enormous penstocks, providing San Francisco with hydroelectric power. The new Priest Grade Road descends 4.5 miles through languid, elevation-losing loops; the brave can duck off just before the new road commences and take the Old Priest Grade Road. It elegantly descends an identical elevation drop in less than 2 road miles—"steeper than a cow's face," as a friend once said. There is no way to avoid using brakes, but shift into the lowest gear possible for a taste of what driving in California used to be like before roads were homogenized for the benefit of truckers and motor-home drivers.

If the road draining tourists from Yosemite is too crowded, another way to Highway 49 beckons. Within 14 miles of the park entrance is *Buck Meadows,* a 2 mile-long sweep of grass, accented with a handsome barn, and sporting the usual convenience stores to the right. A sign at the foot of the meadow points left to the Buck Meadows Road to *Coulterville.* This small byway, which descends through chaparral and forest, connects 6 miles later with road J20. Follow this, the Greeley Hill Road, west to Coulterville; road and town are worth the extra time. So, too, is the tranquility. Coulterville sits among sagebrush and chaparral. Centerpiece to the town is the Jeffery Hotel with its three-foot-thick walls (still offering a bar and renovated lodging), shaded by black locust, catalpa, and umbrella trees—the sole legacy of more than 1,000 Chinese miners who called Coulterville home during the Gold Rush era. Stepping into the hotel bar (the cold beer is good; the food is anyone's guess) is a pace through time. Although the signs are neon, photographs of Coulterville's "Western Gunfighters Rendezvous" are promise of the Gold Country's imaginative mix of historic preservation and surreality. "The past," William Faulkner once wrote, "is never dead. It isn't even past."

Meet Highway 49 at Coulterville and travel north. Before reaching the junction with Highway 120 (where travelers down the Priest Grade Road breathe a sigh of relief on hitting the flats), note the small reservoir below the city of San Francisco's Moccasin Powerhouse. More impressive than the pond are four huge pipes dropping down the hillside to the powerhouse, carrying water from Hetch Hetchy to the turbines. At Moccasin, Highways 120 and 49 join and together move north, almost immediately crossing over outer legs of Don Pedro Reservoir, a lower-elevation dam that reimpounds the Tuolumne River and is a favorite recreational site for San Joaquin and Sacramento valley residents.

Chinese Camp to Angels Camp

Oaks dot the countryside as Highway 49 (it envelops various other numbers in its northerly passage) runs along the edge of the Sierra Nevada piedmont. From early summer until January, the hills are golden, offering an alternative explanation for California's "Golden State" nickname. The road rolls with the hills, showing the oak woodland's advantages as cattle and sheep-grazing country. This is still the southern end of the Mother Lode, where gold-bearing placer gravels, washed loose from country rock by millennia of moving water, were hard to find. More common around Coulterville, Mariposa, and this part of the Gold Country was vein-borne gold, removed from the ground with costly tunneling, dynamite, and sheer brute force. Every possible gold-mining technique has been tried somewhere in the foothill country: placer mining, separating loose gold from stream gravels, got everything going; hard rock mining—burrowing into the earth to remove ore from its country-rock veins; hydraulic mining, in which a high-pressure stream of water was directed against a hillside removing massive amounts of soil and presumably gold; and finally, dredge mining that recultivated the spoils or remnants left from other mining processes—its signature seemingly endless echelons of piled dirt and gravel as if the earth itself was abruptly corrugated.

Ever-present oaks (California has seventeen different species) are a reminder of the efficient economy California's native Indians developed long before the Spanish and Mexican presence began in 1769. Not until the early nineteenth century was anything more than an exploratory expedition launched into the interior of California; many of the Indian residents of the Central Valley were undisturbed well into the 1800s, and the "last wild Indian" of California, Ishi (made famous by Theodora Kroeber's book, *Ishi in Two Worlds*) emerged from the northern gold country near Oroville in 1911. The California Indians probably numbered more than 400,000 at their peak population. They were unmistakably California's native population. But others have arrived since, establishing a firm enough material imprint on the land so that they too are clearly a native part of California's changing citizenry.

Chinese Camp, which lies less than 10 miles up Highway 49 from Moccasin, in its name rightly commemorates two sides of California life—the mining camp and the Chinese contribution to the Gold Rush. Impressive is the ornamental upturned corners of the town's handsome, vaguely Asian-styled Chinese Camp school, which lies just a bit west of the main settlement. The streets are crumbled, a legacy of alternating searing summer heat and rather cool nights; the winters, too, can be frigid. Best about Chinese Camp is the casualness with which its legacy is taken. Chinese-planted locust trees shade houses, and trees of heaven sprout late springtime color. In the early 1850s, Chinese Camp was home to more than 5,000 Chinese miners. They exploited the nearby massive table mountain formation, a mass of Tertiary basalt that flowed down ancient river channels and thanks to erosion and relief inversion now towers well above the surroundings. The basalt cover kept a vast deposit of ancient stream gravels in place, providing miners with rich pickings. Many of the Chinese argonauts organized into tongs, fraternal organizations, and in one of the most famous (and briefest) tong wars in California history, the Sam Yap and Yan Wo tongs engaged one another in 1856, for reasons now

School for children of miners at Chinese Camp. Photograph by Paul F. Starrs.

scarcely remembered. Yankee miners were hired to instruct the tong members in weapon use. Casualties totaled four, and the conflict soon faded.

There isn't anything to do in Chinese Camp but look and think; what's most pleasant about this and other foothill towns is exercising the imagination to conceive what it must have been like long ago. Chinese Camp offers just that freedom.

Returning to Highway 49, drive through Jamestown (near the Humbug mine, which produced nuggets as large as hens' eggs) and Sonora, and follow the Gold Rush Highway toward Columbia. *Sonora* (the Tuolumne County seat) was originally settled by Mexican miners from Sonora, Mexico. Sonora attracted more than five thousand new inhabitants during the first year of gold mining. Legend has it that more than 40 million dollars worth of gold was dug up within city limits. The town still reflects the opulence of its boom-town days. If another small detour is sought, see Columbia

itself. Each has its charm. Heavily renovated, *Columbia* is officially a State Historic Park, restored and maintained in something
vaguely like its nineteenth-century status. Once the "Gem of the
Southern Mines," Columbia was notoriously wicked and fabu-
lously rich. Some small evidence of that survives. The City Hotel
offers lodging and elegant meals, thanks to a culinary institute
associated with the hotel. For those whose imagination works best
with people wandering around in "authentic" garb and who enjoy
stores selling sarsparilla and stick candy, this is the place.

Decent food is for the asking at *Angels Camp,* a paragon among
Gold Country tourist stops. During its heyday, the mining ma-
chinery in Angels Camp ran so loud and ceaselessly around the
clock that townsfolk, it's said, could never sleep. There's plenty to
gape at in this town that embraces a largely fictive identity bor-
rowed from Mark Twain's story about a jumping frog contest;
Angels Camp is up front about its campiness. Every May, a
reenactment of the jumping frog contest is held, and entrants (and
onlookers) converge from far and wide. In the late 1980s, San
Francisco newspapers and even the national press entered the fray
when a Seattle resident proposed entering a brobdignagian African
giant frog in the contest, weighing in at more than a dozen pounds.
After nearly interminable debate about rules ("Rules??" some
questioned . . . in Twain's contest, force-feeding a frog buckshot
helped moderate its performance), a special, but separate, contest
was held. Gridlock seizes the streets during any Angels Camp
tourist event, and that holds true through significant parts of the
summer (the town is three city blocks long, so it clears in good
order).

For the more adventurous, a drive up Highway 4 (it goes east
from a bridge in the middle of town) deposits you after about 8
miles in *Murphys,* named after the brothers John and Daniel who
were the first to discover gold there. The Murphys Hotel is usually
in business, and sports a nice photograph collection. The town
itself was virtually on top of some of the richest diggings in
California; claims in Murphys were limited to a block 8 feet
square. Today, Murphys' wealth is atmosphere and a fine contrast
to Angels Camp, with horsehair sofas, gaslamps, a credible Old

Timer's Museum, scattered mining equipment, and dignity. Albert A. Michelson, born in Murphys, won a 1907 Nobel Prize in physics for his experiments examining the speed of light. It is easy to return to Highway 49 by way of the Murphys Grade, joining the highway at Altaville, a mile north of Angels Camp.

Mokelumne Hill to Coloma

Driving eats up the road north of Angels Camp. Small creeks, accompanied by "Narrow Bridge" warnings disappear quickly behind as tires whiz on the pavement: San Domingo Creek, Indian Creek, Calaveras Creek, Murray Creek. The town of *San Andreas,* on Highway 49 14 miles north of Angels Camp, is a booming foothill community. Like Jackson, Placerville, and Auburn, which also lie ahead on the Gold Rush route, San Andreas is enlarging by leaps and bounds, stuffed with former residents of Sacramento Valley cities who are choosing to live farther from work so they can own a rustic piece of the California countryside, ideally a "ranchette" composed of a house with attached parcel of five acres or so of land. Nothing much is done with the land. Perhaps a few calves, a horse, or sheep graze there, or maybe a llama (exotic animals are a modest rage). This subdivision of once-large foothill properties is simply an answer to demand, and the demand is unending, judging from the proliferating "For Sale" and realty signs in towns of any size. Change may be inevitable, but one thing that's notable about the foothill country is that it successfully turned its back on change for almost a century.

Such dissection is human and contemporary. Geologically, of course, the Sierra Nevada and its foothills are also a topsy-turvy world, much segmented by rivers that changed courses seemingly constantly—but only on a geologic time scale. Near Mokelumne Hill, for example, at least eight different stream channels, all dating from the Tertiary, are identifiable. Finding, tracing, and exploiting these channels—since each carried stream gravels where placer gold might be found—was the monumental study of miners, the most avid of amateur geologists. Their passing grade was a collection of dust or nuggets, although few were the miners who laid claim to even a small fortune.

Mokelumne Hill—it goes by "Mok" Hill—lies between San Andreas and Jackson. Off Highway 49 (barely—about 100 yards), Mok Hill lives with the archetypical notoriety of a former boom town. Supposedly, at one point in the town's history, one man was killed each weekend for seventeen straight weeks. Streets wind back and forth; it's easy to get lost but no big deal, and circling two or three times around the town doesn't in the least diminish its attractiveness. The I.O.O.F. Hall at the foot of Main Street, a fraternal organization metal-shuttered against explosive fires that were the bane of mining towns, is one of California's first three-story buildings; it is shaded during the warm months by ailanthus trees, planted (as was usually the case) by Chinese residents. Several churches survive in Mok Hill, along with a respectable permanent population that occupies crumbling, period housing, rebuilt Victorians, or sometimes mobile homes backed into a driveway. The Ledger Hotel, originally built in 1856 as the Hotel de l'Europe, is renovated and has honest pretensions; its beer is cold.

Highway 49 continues north through Jackson, Sutter Creek, Plymouth, El Dorado, and Placerville. For those with special interest in California Indians, not far out of Jackson (11 miles east on Highway 88) is *Indian Grinding Rock State Park.* At the pine–oak transition, the park site was favored by many of the foothill Indians as a gathering site, and the granite rock for which the park is named has hundreds of round holes worn in it from the action of pestles, reducing seeds, acorns, and plants to the consistency of meal or flour. If geology is more to taste, the town of *Ione* (10 miles west of Jackson) sits atop a patch of Cenozoic history. The clays, colored a deep red, include layers of laterite and occasionally some lignite, a reminder that the climate and vegetation of California were once tropical and heavily forested, perhaps on a par with tropical Mexico of today. Outcrops of the Ione Formation often sprout fifteen-passenger vans, unloading geology students paying homage to an older California.

Placerville, where Highway 49 meets Highway 50, is a modern-day boom town that soon may be inseparable from Sacramento, 39 miles to the west. Highway 50 is a two-lane road 20 miles east of Placerville, and not sweet driving in snow, ice, or rain. The winding streets and old houses of Old Town Placerville, a community within a community, contrast with newly built thoroughfares and sanitized housing tracts. The seat of government for El Dorado County, Placerville was a strategic point on both the Overland Trail to California and the road to Coloma. The Overland Mail, the Pony Express, and the first telegraph line passed through Placerville; its notable early citizens included Mark Hopkins (of railroad fame), Philip D. Armour (of meatpacking note), and John Studebaker, who built wheelbarrows for miners before starting in the automobile business.

Coloma (named for a Maidu Indian settlement) is 6.5 miles north of Placerville on Highway 49 (Placerville is always under construction, so keep your eyes on the sign, or ask how to get out of town). It all began at Coloma, on the South Fork of the American River. In January of 1849, James Marshall noticed a few flecks of gold in the millrace of John Sutter's sawmill. This wasn't the first discovery of gold in California—in 1842, gold was found and worked in Placeritas Canyon near Los Angeles. This one, however, galvanized the world. By summer, 2,000 miners had arrived, by the next year 10,000, and most '49ers came to Coloma first, although its deposits were virtually gone by 1852. Food was impossibly expensive, and pick and shovels cost $50. James Marshall, who made the discovery, died utterly impoverished in 1885 and was buried on a nearby hill.

While such tales of woe were common in California, there are equivalent accounts of success, not only in California during the Gold Rush, but after and still today. Coloma, and the modest and handsome state historic park there, is both symbol and reality. A re-creation of the sawmill, moved from its river site to nearby dry land, is one exhibit at the park. Displays of various other mining technologies, and explanations of how they functioned, capture some of the sense of excitement, innovation, and frenzy that touched this small, obscure part of California and turned it into a focus of world attention. Reprints from the San Francisco-based *Mining and Scientific Press* suggest how much the apple of the world's eye

California was for a time. In its day the *Press* (which lasted well into this century) was the best means of getting information about mineral exploitation, new technologies, and mine yields. It was the trade journal par excellence. Coloma State Historic Park is understated, perfectly adequate to its task, and the picnic tables sit in ample shade where visitors during the warmer months can relax.

Coloma is the core of the Gold Country. It is a great tribute to the state of California that the discovery site isn't surrounded by houses or modern buildings. The Coloma Valley, split by the American River, is as good an example of what the discovery of gold meant to California as might be asked for. It remains tasteful and pleasant—it's the rest of the state, with 30 million residents in 1990, that has arrived in the meantime. Seeing Coloma makes Highway 49 to this point, and the towns yet to come, more understandable.

Deep-Rock and Hydraulic Mining: Nevada City and Vicinity

A couple of dozen miles north of Coloma on Highway 49 is *Auburn,* and the junction with Interstate 80. For those who like interstate highways, I-80 is a type specimen: at least four lanes, generally made of super-durable concrete, an almost all-weather road (it closes rarely and opens as soon as formidable highway department technology can open Donner Pass), and carefully patrolled for traffic safety. If the weather is bad, take advantage of Donner Pass. Nearly everyone in northern California knows by heart the phrase "Chain restrictions are in effect from . . . [Baxter or Colfax or Applegate or Auburn]"; the Interstate is kept open when every other pass closes. Just be prepared to put chains on your tires—"chain monkeys" will happily rent you a set and put them on for you.

George R. Stewart, novelist and long-time professor of English at the University of California, had a better command than anyone else of this part of California. What Stewart admired most were the small details of the Sierra Nevada foothills: names that spoke of their origins, the greedy and savage alteration of the natural envi-

THE DONNER PARTY

> I have not rote you half of the truble we have had but I have rote you anuf to let you now that you dont now what trouble is . . . we have left everything but i dont cair for that we have got throw with our lives but dont let this letter [disharten] anybody never take no cutofs and hury along as fast as you can. —Letter by twelve-year-old Virgina Reed, 16 May 1847, three months after her rescue (in Stewart, p. 287).

The Donner Party was neither the first nor the largest organized party of emigrants to set out for California in the 1840s, and none of their members would go on to shape the destiny of the state directly. Yet, their deeds are easily more notorious and noteworthy than those of any other such group. The Donner Party story provides two lessons: The Sierra Nevada was a formidable barrier—or shield, depending on point of view—isolating the inhabitable parts of the state; and resettling in California required enormous amounts of determination and will. In their own way, both are still true today.

The California Trail was declared open and feasible in 1845, inspiring a great migration the following year. On 28 July 1846, one group of midwestern emigrants led by Illinois farmer, George Donner, arrived in good shape at Fort Bridger in what is now southwestern Wyoming. In all, there were eighty-seven persons, including forty-three children, spread among twenty-three wagons bound for Sacramento before winter.

Their troubles began just across the Continental Divide. After taking a full month to get near Salt Lake City, Donner chose an ill-conceived cutoff around the Great Salt Lake. Moving at a snail's pace, the party quickly became the rear guard in the emigrant train of 1846. Their luck was not much better crossing Nevada: The first of the party arrived late and exhausted at the eastern front of the imposing Sierra edifice on October 19. The group itself was stretched out over two days' travel.

The party reached Truckee (now Donner) Lake at the same time as the first heavy snowfall that year. The first of no fewer than six different attempts to scale the summit failed on November 1 due to a raging storm. Expecting that the snow would melt between storms as it did in Illinois, the party was not immediately panic-stricken and returned to the lake to wait for another opportunity. The storm on the pass continued for ten more days, and the group soon realized they were in trouble.

In preparation for a long winter, cabins were hastily built at Donner Lake and at a site farther back on the trail. Food was in short supply, however, and they found hunting impossible. Other attempts at the pass were made, first by pushing and pulling wagons and emaciated livestock up through crevices in the craggy cliffs, and then finally, on foot. Each succeeding effort was met with colder temperatures and deeper snows, forcing retreat back down the mountain. On December 13, the snow was already 8 feet deep at camp, and the remaining livestock had disappeared. The first death was recorded on December 15.

On December 18, fourteen of the party—the group later called The Forlorn Hope—with provisions for six days, made it over the pass only to wander south into still more rugged terrain. Thirty-three days passed before half of them arrived at a friendly Indian camp, having been compelled to eat the bodies of their comrades who had died along the way.

Meanwhile, more had died back at the mountain camps, and on January 4 through 8, Virginia Reed, her mother, and two other adults made yet another desperate attempt at the pass on foot. Exhausted and starving, they retreated back to camp to find the snow level now at 13 feet.

The first relief party from Sutter's Fort finally reached the remnants of the Donner Party on February 18, and attempted to get out with twenty-two survivors. However, predators had destroyed food caches left for the return, and this group too was reduced to starvation before meeting a second relief party. This second group returned to the camps on March 1 to find more dead and clear evidence of cannibalism. Weak, emaciated immigrants and rescuers tried to get out, foundered in heavy snows, and again starved and resorted to cannibalism.

There were few left alive in the cabins by the time a third relief party arrived in mid-March 1847. Only four were strong enough to make it out; the rest were left behind. The fourth and final relief effort went out in the spring and found all of them dead, except one man who appeared content and well-nourished. In sum, thirty-four of the eighty-seven who began the journey had perished in the Sierra mountain camps or on Sierra trails, and in so doing contributed to the survival of the others. Today, the huge Pioneer Monument standing at Donner Lake just off Interstate 80 testifies to this heroic tragedy, and to the emigrants' fight to settle in California.

ronment from any of a variety of causes, and the welter of human stories that attached to places and activities and weather and everything that goes with them. Perhaps nowhere are these stories so tightly concentrated, and so dramatic on the land, as in the area around Nevada City.

From Auburn (notable for handsome old architecture) to *Grass Valley* it's 24 miles on Highway 49. Grass Valley and Nevada City are usually twinned, but they are an odd couple: Grass Valley keeps up with the times, while Nevada City is reaching for the 1990s with quaintness in mind. Grass Valley's mines kept producing well into the twentieth century. Now it has nearly a dozen museums. The Nevada County Historical Mining Museum and the

Empire Mine State Historic Park are two worth seeing (both keep daylight hours, but check with the Chamber of Commerce for precise hours). Grass Valley is psychically aligned with Sacramento and the valley's happenings, thanks in part to a large number of refugees moving to Nevada County in the last decade. Take from Grass Valley whatever you want; a number of visitors take a good look, shift back into gear and travel the next 4 miles into Nevada City.

Maybe *Nevada City* doesn't really have the highest concentration of bed-and-breakfast lodgings per capita in California—it just seems that way. The exquisitely restored Victorian houses are perfect for putting up guests from other parts, and a modest walk through town shows why it is a great destination. The grand National Hotel was built in 1854 and is a working hotel, well worth a close look. Search out the continuation of Highway 49 as it leaves to the west of Nevada City. After leaving the freeway section of 49, watch for the Forest Service Ranger District headquarters on the right. Almost immediately after the headquarters is a sign for the North Bloomfield Road, and take it.

Malakoff to Truckee

Nearly 20 miles up the North Bloomfield Road are the *Malakoff Diggins,* site of another State Historic Park. ("Diggins" is Malakoff's quaint, but accepted, spelling of "Diggings," the traditional description of these mining operations.) The road is not for the timorous, and just after crossing the South Fork of the Yuba River, it turns to a gravel roadbed. (A good road, but gravel; if the weather is bad ask before turning off at the Forest Service office, where they will know travel conditions.) Going to the Malakoff Diggins is a great excursion, either as part of a continuing trip or as an out-and-back journey from Nevada City.

MALAKOFF DIGGINS

In a couple of decades the San Juan Ridge geological formation was battered as perhaps no place else on earth. At Cherokee, North

MINING TECHNOLOGY

The Malakoff Diggins was an end product of hydraulic mining. The salient principle of "hydraulicking" is simple. Placer mining involves the collection of auriferous gravels, dirt, and sand; sifting them in water so any gold that might be included in the debris will fall to the bottom; and then keeping the gold. Everything not gold is discarded. Whether the equipment used is miner's pan, rocker cradle, or sluice, this is the basis of placer exploitation. The most ambitious scale of placer gold-mining is hydraulic mining. Control of financial capital, land, and water is essential.

First, water is diverted far upstream, usually into a wooden flume (hydraulicking placed great strains on the western Sierra forests). The water is moved as nearly as possible on a contour, so no elevation is lost. As the mine site nears, the water is diverted from flume into a pipeline that narrows as it drops, with the pipe growing more heavily reinforced and narrower as the water descends and pressure builds. Finally, the liquid is released through a cannon-like nozzle known as a hydraulic monitor, invented in California in 1853. The opening ranges from 4 to 12 inches, and the stream water—expelled by a force that increases at a rate of half a pound-per-square-inch for each foot of elevation drop—spews out at pressures in excess of 300 pounds per square inch, or enough to throw a solid stream of water in 400-foot arc. The hydraulic monitor, carefully counterbalanced so that a skilled operator could move it back and forth with minimal effort, blasts away at an opposite hillside and the "debris" (an entire hill itself) washes downstream where gold is separated from the soil and gravel slurry in a series of sluices.

The technological marvel that was hydraulic mining in California spelled doom for the independent '49er. With the sea of capital (often foreign in origin) needed to acquire mine properties, hire workers, and build flumes that moved water to the diggings, the rise of hydraulic mining in the late 1850s closed the frontier era of prospectors who traveled from claim to claim with burro and pick axe. North Bloomfield, site of the Malakoff Diggins, was probably the biggest hydraulic mining operation in the world from 1866 to 1884, when legal action finally shut down hydraulic mining throughout the northern mines because silt moving downstream made river passage impossible, caused enormous floods in the Sacramento Valley, and threatened an increasingly lucrative farming industry.

Columbia, Lake City, at Malakoff, farther south at the Alpha and Omega mines, and at numerous sites between, hydraulicking was effectively practiced. That devastation is readily apparent at the Malakoff Diggins. Tough roads and ignorance about the scale of hydraulic mining keep visitor numbers down at the park—the pilgrimage is nonetheless worth making. Scars are there for the seeing; they are impressive, but somehow not depressing. Manzanita and madrone, the generic Sierra Nevada foothill vegetation, dot the hillsides and cover some of the debris, but the sharp gash is a testament to earth moving. The silence is unnerving, and talking above a whisper feels like a sacrilege. In *North Bloomfield* a couple of miles up the road from the Malakoff site, are simple cottages and the former mine office, where gold was reduced to bars for shipment to the San Francisco mint. The largest bar, the 1939 *Writers' Project Guide to California* claims, weighed a quarter of a ton.

From the Diggins, the choices are several. Returning to Nevada City and connecting there with Highway 20 is a simple backtrack. A sufficiently detailed map will show how to return by different routes, but the roads are labyrinthine, and best not attempted without a map. For the brave, the Relief Hill Road (labeled along its course as Road 36) heads south from the middle of North Bloomfield. This is a little bit better than a logging road (and does

Mining scars in the foothills of the Sierra Nevada. Photograph by Paul F. Starrs.

appear on good maps) and should never be tried in questionable weather. The road is named because a second group, sent out from Nevada City to attempt a rescue of the snow-stranded Donner party near present-day Donner Lake, supposedly met with refugees and their rescuers on Relief Hill. The route is back-roads California at its best: not an especially challenging drive, but an exhilarating 10-mile cut through time. When I last drove it, I saw endless deer and a large fox; a friend nearly ran into a bear some years back. The Alpha and Omega diggings, relicts of the southernmost San Juan Ridge hydraulic mines, lie just off the Relief Hill Road, which connects with the Yuba River almost exactly at Washington. Rarely does two-lane pavement look quite so good.

WASHINGTON

The town of *Washington* is like a bizarre Shangri-La or Brigadoon caught in a time warp. Less than 6 miles off Highway 20 (also easily reached directly from Nevada City), Washington is on the Yuba River, a hamlet lit up at night like a Swiss village, with barking dogs and a crisp aura of wood smoke scenting the night air. A couple of hotels/bars offer lodging and drink; neither is bad. It's the residents, as is so often the case in isolated foothill towns, who give Washington a spicy flavor. Parts of the Sierra Nevada that are too inaccessible for commuters to Sacramento, too inconveniently small to attract urban refugees, remain the domain of people who make a deliberate choice to escape the mainstream.

Washington is the rural counterpart of Haight-Ashbury. In many small towns, from Nevada City north, the vehicle of choice is an older pickup truck; women wear long skirts and braided hair; men look like prospectors or motorcycle outlaws with long beards and mustaches and hair pulled into pony tails, and Levis to accompany leather vests (shirts optional in summer). The alternative economies of California play important roles here—cutting wood and selling it, working in handicrafts, taking in work from one another, barter, living on unemployment, or growing dope. Don't ask, if someone drives out of the forest with a truckload of something under a tarp. You weren't meant to be there. The fine end of the scale is North San Juan, where writers and teachers and loggers and other people have formed a remarkable sort of community, inspired in part by poet Gary Snyder, who lives there. Washington is another sort of place: At the Yuba House, the beer on draft is always Budweiser, and you get it in pints or quarts served in Mason jars. To discourage bikers and gang members, a sign above one establishment says "No Colors," and in the evening you can walk the fifty steps through town and lie down on the middle of the road without worrying about traffic. (Make sure no one is leaving one of the bars, though.) Washington is worth the stop. Nothing will change there; you can just about count on it. The road leaving Washington joins Highway 20; turn east to connect with Interstate 80.

WASHINGTON TO TRUCKEE

A dozen miles up Highway 20 from the detour to Washington, keep an eye on the left side of the road. Good reflexes will allow you to stop right after seeing a substantial wooden structure, crossing under the highway. This is the *South Yuba Canal,* or flume, a surviving example of the transportation system used to bring water to hydraulic mines; this flume is still in use today as the water supply of Nevada City. For once, in California, there are no "Keep Off" signs, and the catwalk running down the middle of the flume makes a pleasant stroll above running water eminently possible. Flumes that provide town drinking water are still patrolled in winter by "ditch walkers," responsible for breaking the ice, removing jams, and maintaining the flow. A warren of these flumes once criss-crossed the Yuba-Bear River drainages; relatively few are still in use, although there is a surfeit of small dams high in the Sierra, providing hydroelectricity for Pacific Gas and Electric, the huge California utility whose great vision was to recognize the value of "free" power, there for the asking with only a bit of dam work and some pipeline laying.

Once back on Interstate 80 (follow signs to Truckee or Reno), stay on the pavement and rumble east on the nation's most important all-weather pass across the Sierra Nevada. The concrete is rough, gouged by tens of thousands of slapping tire-chains. The summit, elevation 7,239 feet, is almost anticlimactic. Keeping the

South Yuba Canal. Photograph by Paul F. Starrs.

View of Lake Tahoe. Photograph by Adrian Atwater, courtesy of the University of Nevada Press.

interstate open isn't always easy—Blue Canyon, just off the road near Emigrant Gap—holds a number of lower-48 state snowfall records. The railroad tracks lie south of the Interstate, and for a considerable part of their traverse, trains to this day move from tunnels carved in the Sierra into snowsheds, huge covered sections of track, that make a winter crossing by train possible but not guaranteed. As the road descends, *Donner Lake* also lies to the south. The Donner Party attempted to cross the Sierra in late fall, were caught in several severe snowstorms, and ran short of food. While some cannibalism followed, the details are subject to debate. A monument to wiser, more provident, or at least more successful immigrants stands at Donner Memorial State Park, at the eastern end of Donner Lake, easily reached from the Interstate.

Truckee, once a toothsome railroad town, has gone to kitsch. The railroad still bisects the town, and anyone with a powerful desire to see *Lake Tahoe* can follow one of the two roads (Highway 89 or 267) to the lake. Aside from the railroad tracks, and the decisive thud of trains passing through town, the best thing about Truckee is the advertisement coming into town for the "Ponderosa Ranch," where the television series "Bonanza" was filmed, near Incline Village at the northeast corner of Lake Tahoe. There are lots of knickknacks to buy in Truckee, and food, gas, and T-shirts. With Nevada and Reno ahead, wiser travelers practice thrift.

Reno, Nevada

Interstate 80 from Truckee to *Reno* is powerhouse pavement. Everyone who drives seems to be rehearsing for a run at the land-speed record, swooping downhill through wide turns and six lanes. Ahead lies the Nevada border, and it used to be that the first sign of Nevada civilization was steadily drying hillsides, dotted with sagebrush, and then abruptly, a scattering of handsome meadows making use of Truckee River water that descends with the Interstate from Lake Tahoe. Billboards and neon now precede the pastures, appropriately enough. *Verdi,* once a lumber camp, is now an extension of Greater Reno, 11 miles away. Houses began ap-

Reno, the biggest little city in the world. Photograph by Paul F. Starrs.

pearing there in the early 1980s; now Verdi is a metropolis. So, too, is Reno.

Anyone who's been to Reno before remembers the sign, "Reno, The Biggest Little City in the World." That now seems overly nostalgic. With exploding suburbs, air contamination problems, and skyrocketing water bills, hardly anything is little about Reno anymore. Nonetheless, the center of town is still along Virginia Street, the name taken by Highway 395 as it cuts a swathe through the city. That's where the "Reno" sign was, and still is. The impressive big hotels are increasingly being built farther from the central business district. Reno remains a city without an attitude. Amtrak trains continue to move through the middle of Reno. Just north of the Interstate is the University of Nevada, shifted to Reno from Elko in 1889 in a move that led residents of other communities to sneer that Reno pretended to be the "Athens of Nevada." Basque restaurants and hotels are not far from the train tracks, reminding those in the know that even today Nevada has a large Basque population, although few Basque-Americans embrace the shepherding trade that brought their parents or grandparents to the American West. Now someone with a Basque name is more likely to own a casino, teach in the University of Nevada's Basque Studies program, or be a well-known politician. Things change, and for Nevadans of Basque descent, change has generally been for the better. There is nothing especially spectacular or singular about Reno. Even the frenetic energy of the gambling district is unassuming: They want your money but are willing to make losing it fun.

In 1900, Reno was a dive; a brazen community with a boom-and-bust population cycle that generally followed the fortunes of the railroads. The Reno of gambling, divorce, shady politics, and fast-living California visitors was hardly a gleam in anyone's imagination. It was simply another tough railroad town. In 1904, Reno was so raffish that, according to Nevadan David Toll, the president of the University of Nevada restricted students to campus grounds to protect their virtue. About that time, the Nevada divorce was invented by a lawyer who specialized in sundering celebrity marriages. Nevada had a relatively short six-month residency requirement and a long list of reasons that could be cited as grounds for divorce. The boom was on, especially since Reno and surrounding towns developed an assortment of places for divorce-minded visitors to stay. A six-month residence on a Nevada dude ranch was enough to make almost anyone a fan of the Silver State. An ebb and flow in gaming laws brought Reno increasing fame as a gambling center. Yet Nevada, and Reno specifically, is in many other ways quite puritanical. People who live there have strict standards and don't countenance too much wildness, except where parting tourists from their lucre is concerned.

The best thing to do in Reno is park near Virginia Street and walk. Within two blocks of either side of Virginia are tree-lined streets, residential neighborhoods with fine and not-so-fine houses, and a more balanced sense of what Reno is really like than anyone might expect on first seeing its glowing self-endorsement. To leave Reno (and more and more people are coming to stay), drive south on Highway 395. If you have no desire to see any of the downtown, there is a freeway version of 395 to the east of the business route. It's quicker, if lacking in substance. The road takes traffic speedily south of Reno, except at rush hour. Mustangs (feral horses) graze on Bureau of Land Management (federal government) ranges within 20 miles of the city. An introduced and immensely destructive pest, they're also undeniably exciting to see for the first time.

On and Off 395: Virginia City, Carson City, and Genoa

At last check, nearly 87 percent of the state of Nevada was owned by one government entity or another. Most of the state is public land and administered following rules generally developed in Washington, D.C., and not in Nevada. Local residents periodically express unhappiness about this situation, and actions like the abortive "Sagebrush Rebellion" of the late 1970s are attempts to balance local needs (or desires, anyway) with federal ownership. Every western state has somewhat the same story; the voice of

Nevada, with the largest proportion of federal land and until recently a history of very senior and powerful senators, carries particular weight. But Nevada is also a different place in the 1990s than it was even twenty years ago. The state is highly urbanized; only California has a higher ratio of city-to-rural residents. The population of Nevada is growing rapidly, most of all in Las Vegas, Reno, and the Carson Valley, which includes the state capital of Carson City and lies south of Reno. Concerns of an older generation of Nevadans, worried about too many wild horses, new subdivisions, costly road projects, homeless people, or not enough respect for the traditions of a state that was created as an agricultural and mining empire, go to the back burner. With far more than twice as many Nevadans in the state in 1990 as 1970, growth and its costs are on everyone's mind. There are still lots of fine sights to see—they just have to be kept in context.

About 150 miles south of Reno, following Highway 395 all the way, is Mono Lake. While Highway 395 is a decent road, it can be congested, and there are always alternatives. Nevada's equivalent of Coloma is Virginia City, and it was silver to California's gold. Ten miles south of central Reno is a turnoff for Virginia City. Fifteen miles later, after a climb into the Virginia Range, is *Virginia City*. Like the mountains, Virginia City itself is a sight to behold.

The mountains have a simple enough story. All of Nevada is being stretched, east and west. The Sierra Nevada forms the western boundary of the Great Basin and the Wasatch Front of Utah (looming above Salt Lake City) the eastern fringe. Everything between is under considerable tectonic stress, part of a phenomenon that is observed but not entirely understood. The Earth's mantle is actually thinner across Nevada than almost any place else in the world; mountain ranges, rising constantly, are uplifted; hydrothermal activity is common, likewise earthquakes, and this weakened skin of the Earth is prone to all sorts of unlikely events. The pulling action across Nevada is geologically well-established, and its clearest physical manifestation is more than 120 mountain ranges that invariably trend north–south. Nineteenth-century geologist Clarence Dutton, in a deservedly famous phrase, described

Virginia City, Nevada. Photograph courtesy of the University of Nevada Press.

Nevada's mountains as looking like "an army of caterpillars, marching toward Mexico." The mountains and valleys have a classic form, generally referred to as Basin and Range structure. Much of Nevada is already high desert—the valley floors are frequently a mile above sea level. Rising next to the valleys are ever-present small mountain ranges, typically standing a mile above the valley below. These abrupt mountains provide summer grazing forage for wildlife and livestock, water for ranches and crops, and an up-and-down drive for anyone crossing central Nevada, and actually are ecological islands where the theory of island biogeography is sometimes tested.

A by-product of this active geological history is an enormous amount of mineral deposition in Nevada's fractured rocks. Some

*Miner's Union Hall and Piper's Opera House, Virginia City, Nevada.
Photograph courtesy of the University of Nevada Press.*

of the minerals are of no account, exciting for mineralogists alone. But at other times, gold, silver, and copper were left behind. Near Virginia City, in 1859, came discoveries of gold and silver, after a decade of uncertainty about local mineral resources. Gold was found first, but before long the unending supply of "black stuff" that gold miners cursed turned out to be silver, and the boom was on. The townsite in the Virginia Range could be hellish or unbearably cold. That is still the case today, and sagebrush-covered hillsides don't deflect the course of the wind at all. If tourist flavor is strong, that's always been the case; Virginia City provided a number of Californians, including George Hearst and Adolph Sutro, with vast fortunes.

Virginia City is today a place of contrasts. The St. Mary's of the Mountains church, a block from the main street, is quiet and pretty, supported by visitors and by the 700 or so people who live in and around Virginia City. Houses are small and well-maintained. The main street is an unmitigated zoo, with every known form of enticement vying for tourist money. The Bucket of Blood Saloon is a favorite; the Crystal Saloon another; and on the hill behind town a huge painted "V" broadcasts the city's identity, in case anyone was unsure. Painted-up though the buildings are, inside most echo 130 years of history. Look for historic photographs or explanations of engineering marvels like the Sutro Tunnel, designed to connect with deep shafts. Examine schematics of the mines themselves, or facsimile editions of the U.S. Geological Survey reports on the Comstock's mineral prospects. Behind all the "fakelore" for the credulous is substance, and Virginia City has an unending supply of that.

When leaving, the fastest return to Highway 395 is south to Highway 50, then west to 395, about 13 miles total. The back route is more fun—follow the signs to the east for *Six Mile Canyon* (where initial discoveries of silver were made) and plummet down the excellent gravel road. Car drivers on Six Mile Canyon are far less nervous than tourists on the paved roads, and in driving down the canyon a sense of drama develops as first the Virginia, then the Flowery Range falls behind in a trail of dust as you descend onto a shadscale- and sagebrush-pocked plain. Housing appears, and directly ahead is Highway 50; go right through Dayton and connect with Highway 395 in downtown Carson City.

Nevada became a state in 1864. The capital is *Carson City.* The State Library is there (in a restored older building; very nice). So is the Capitol. Downtown is the Nevada State Museum, which includes a well-preserved base relief map of Nevada that is best seen from the second floor. Beyond that, enjoy time in Carson City as an interlude—with almost everyone in Carson City a government employee of some kind or another, or a lawyer, it has aimed its sights low and kept some dignity.

About 5 miles south of Carson City, not long after the junction where Highway 50 joins Highway 395, a road turns off 395 to the right. This is Jacks Valley Road, and it leads gently toward the Sierra Nevada, past working ranches, increasing numbers of "ranchettes" purchased by "California cash buyers" who are moving to Nevada with their home equity, and an assortment of ramshackle outbuildings. West is the Toiyabe National Forest, and the tallest pinnacle in sight is Genoa Peak (9,150 feet). This westernmost part of the Carson Valley is Jacks Valley, and the closest buildings to the mountains are *Genoa,* "birthplace of Nevada."

Genoa is a jewel, and far richer in history than a number of more populated places. Part of the charm of going to Genoa is eliciting its history from residents; but in a nutshell, Genoa was originally known as Mormon Station, explored in the late 1840s by Mormons contemplating a peopling of the State of Deseret, created by a decree of Brigham Young's. Mormon Station was at the westernmost edge of Deseret (the Mormon vision of empire was brought up short by fervent response to the California gold discoveries and the silver strikes in the Nevada Territory). When Mormons were recalled to Salt Lake City in 1857, Mormon Station became Genoa, named after the birthplace of Columbus. A substantial freighting industry developed, climbing up the eastern Sierra face on the Kingsbury Grade (with a 15 degree grade). Nevada's long-printed *Territorial Enterprise* was printed for a time in Genoa, and "Snowshoe Thompson" delivered mail on snowshoes and, later, skis.

A restoration of the original Mormon Stockade forms a nice park, and Genoa is struggling to get the minimal tourist traffic to stop and shop; likely it will be decades, if ever, before anything substantial gets going. Genoa does offer a chance to imbibe at the oldest drinking establishment in Nevada, and a conversation with the bartender can be a mind-expanding experience. The town wears the mantle of its history lightly. Ultimately, it is the valley, the empty road, grazing cattle, the basculating block of the Sierra Nevada towering to the west, and the dignified ambiance that make Genoa extra special. The Genoa Road leads directly back to Highway 395, where cruise controls can be engaged.

Those in a hurry can stay on Highway 395 as it moves back and forth through valleys at the eastern foot of the Sierra Nevada. At *Topaz Junction,* 19 miles south of Gardnerville, is one final detour

Ranchland typical of Smith and Mason valleys. Photograph by Paul F. Starrs.

to entice anyone who wants to see a sleepy Nevada agricultural valley and drive a stunning route along the East Walker River. The travel distance by this route is 4 miles longer than staying on 395. A left turn onto Nevada Highway 208 is a move toward Wellington and then, on Highway 338, to the Smith and Mason valleys. Nothing is insistent about the landscape; for those who like the desert, delight in jackrabbits bolting across a road, find cattle grazing on lush pastures a balm to the eyes, and revel in the scenery of the rural West, this hour-long drive is a great restorative.

Bridgeport is the last substantial town before arriving at Mono Lake, and if you've taken the Smith Valley route, Bridgeport lies to the right when the East Walker River Road joins 395. The Bridgeport Valley is notorious as one of the coldest places in California, but its restaurants are terrific and the vistas more than adequate. County seat of Mono County, California, Bridgeport boasts a courthouse that (astonishingly) houses virtually all official county functions. Mono County is not exactly a booming place, which is fine with most residents.

Highway 395 from Bridgeport to Mono Lake (and Lee Vining, beside the lake) goes past moraines, rushing Sierra Nevada streams, and across Conway Summit, named for a Mono County pioneer. A vista point just south of the summit offers the best road-view of Mono Lake. Less than a mile south of Lee Vining, the Tioga Road descends and links up with Highway 395. As realtors' signs say, "if you lived here, you'd be home now." From Mono Lake, the world of the eastern Sierra is your oyster: The White Mountains lie ahead, if you turn east on Highway 120. The glacial history of the Sierra is south, alongside Highway 395. In Bishop is more of the story of Los Angeles and its ingenious accumulation of water rights for the largest metropolitan region in the United States. Roads go everywhere; they are there for the driving.

Bishop to Death Valley

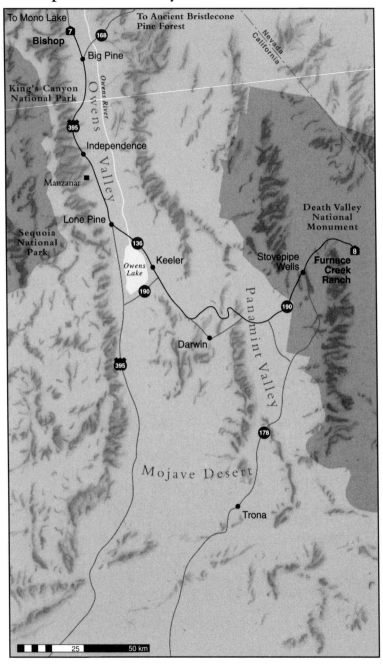

△ *Day Eight*

MONO LAKE TO DEATH VALLEY

Rain Shadow and Ghost Lakes

The Mono Valley landscape is one of sharp contrast and subtle gradation. Coming into the valley from Tioga Pass we see indigo Mono Lake dominating a stark and near-barren basin. There are two islands in the lake, although at very low water levels they lose their integrity. The smaller, northern island is named *Negit* (or, blue-winged goose). The dark coloration of Negit derives from its basaltic cinder cone. A similar cone, Black Point, can be seen on the mainland northwest of Negit. The larger, light-colored island is *Paoha* for sprites of the vapors. Little vegetation grows here other than what clings to the narrow belt of springs and creeks along the base of the Sierras—some Jeffrey pine near the southern margins of the lake, and Great Basin sagebrush spotted over the open drylands.

The Mono Valley is high desert. The elevation of the lake surface is more than 6,000 feet above sea level. Precipitation averages about five inches per year. Winters can be brutally cold and summer afternoons quite hot (although merely balmy relative to Death Valley).

The basin was occupied by the Mono Indians when California became a state. Immigrants were drawn to the valley in search of new pasturelands, opportunities for trapping, and mineral wealth. Just north of Mono Lake were the rich mines of Bodie, and there were lesser mining regions in the canyons leading back into the

Composite photograph of the California-Nevada border. Yosemite National Park and Mono Lake are at top left. The Sierra Nevada range runs diagonally to the left of Owens Valley; the White Mountains, to the right. Dry (light-colored) Owens Lake is at the lower end of Owens Valley. Mount Whitney and the High Sierra are to the left of Owens Lake. China Lake Naval Weapons Center is at bottom right. The city of Bakersfield is at bottom left edge. High-altitude aerial photographs by United States Geological Survey.

Sierra Nevada. Cattle, to feed the population of hungry miners, were grazed extensively where mountain streams led across the valley margins.

Mono Lake

Leaving Lee Vining (named after Leroy Vining, an early rancher), we follow U.S. 395 south.

About 5 miles south of town, take a left on Highway 120 (east) and go 1.5 more miles and watch for the sign to South Tufa Area. Take the road (now a well-graded gravel surface) one mile along a series of ancient raised beaches. Ahead, to the right, we see the tufa pinnacles of *Mono Lake*. Our road ends at a parking area. Look around at these weird reminders of former lake levels. (In summer, there are free guided walks here. Call 619-647-6629 for more information.)

In the vicinity of the parking lot, notice also the raised beaches associated with once-higher lake levels. During the late Pleistocene (ca. 10,000 years BP), Mono Lake was at least 700 feet deeper than it is at present. At its maximum, the lake overflowed the basin and drained south, toward the Owens Valley. The lower series of these raised beaches is relatively modern, representing the drop in lake level associated with the water mining conducted by the Los Angeles Department of Water and Power.

Mono Craters

From the parking area, the route leads directly south to Highway 120, and then left, toward the east. Mono Craters are on the immediate right. At this point we are in the *Inyo National Forest*. This route leads us through several stands of Jeffrey pine rooted in pumice deposits from eruptions in the Mono Craters. There is very little other vegetation along this stretch. The barren ground is largely due to poor soil development, lack of precipitation, and rapid infiltration of moisture through the pumice. After 2 miles, a

TUFA

As you leave U.S. 395 and begin dropping toward Mono Lake, you can see two distinct sets of tufa deposits. The first is near the lakeshore to our left. The second, more extensive deposit is in the distance to your right, and is our immediate destination. The pale spires rising from the water and along the shoreline were once submerged within a much deeper and larger Mono Lake. Evidence of the higher lake levels can be seen in abandoned shorelines at several elevations around the present lake.

The tufa spires are located at the sites of former aquifer-fed springs. These springs brought highly mineralized waters, rich in calcium bicarbonate, into the lake. Before exiting the springs, the water is supersaturated with calcium bicarbonate because of elevated hydrostatic pressure in the groundwater flow. Release of flow into lakes reduces the pressure, causing the carbon dioxide to de-gas from the spring waters. This precipitates the formation of calcium carbonate pinnacles of tufa. In some environments, the depositional processes are enhanced by blue-green algae (cyanobacteria). Tufa is a porous form of travertine, and it is related chemically to the stalagmites and stalactites that form in caverns. During their formation, the spindly towers are buoyed by the water around them. Similar deposition in air can occur, but the results are low, mound-like formations.

Lowering of the lake level, largely due to diversion of water to the Los Angeles aqueduct, has exposed the tufa landscape at Mono Lake. Similar exposures can also result from changes in climate or local drainage, and exposed tufa deposits are frequently used as indicators of climatic change. For example, as the climate of California became drier at the end of the Pleistocene (about 10,000 years ago), many large lakes in the eastern and southern portions of the (present day) state dried up, leaving travertine deposits as indicators of former submersion.

At present, the waters of Mono Lake are quite alkaline (pH exceeds 9.5) because of high evaporation rates from the lake surface and the lack of drainage from the basin. Mineral concentrations have increased through time as salts are left behind during evaporation. At present, dissolved solids are present in concentrations of about 6 percent by weight. The alkaline waters, as a result, have a distinctively slippery feel. Brine shrimp and fly larvae are the only organisms capable of tolerating this brine soup. The latter, in particular, were a delicacy enjoyed by the Monos.

Outside Trona is an extensive field of tufa pinnacles left by the retreat of Searles Lake. The Trona Pinnacles are a registered National Natural Landform, composed of more than 500 individual tufa towers. Close inspection of the tufa there reveals its "melted crystal" structure and the complexity of the depositional processes. Each of the pinnacles had one or more hollow cores through which springs flowed, and some of these cores can be recognized near the tops of individual towers. Lake Searles was once about 640 feet deep, and, of course, all of the pinnacles were submerged.

historical marker on the left side of the road marks the former location of Mono Mills. At this site, timber from the adjacent mountains was milled for use as lumber in the nearby settlements—Bodie, in particular. Turn around and follow Highway 120 back toward U.S. 395.

Driving west on Highway 120, several of the old, raised shorelines of ancient *Lake Russell* can be seen across the pumice flats. They are most noticeable along the eastern edges of the basin. Evidence of extensive alpine glaciation during the Pleistocene can be seen directly to the west. Note especially the U-shaped valleys and the extensive lateral and terminal moraines. Avalanche scars, caused by both rock and snow slides, are also prominent

Tufa near Mono Lake. Photograph by Douglas Sherman.

on the Sierra flanks. The different coloration of the peaks reflects differences in geology, especially between the lighter-colored granites, the dark basalts, and the variegated metamorphosed sedimentaries.

Ahead, you can see the arc of the *Mono Craters*. At least fourteen distinct craters in this series range in age from 1,200 to 60,000 years. The youngest in the series is Panum (or lake) Crater, the only cone appearing on the north (right) side of Highway 120. The crater is composed of a tephra ring about 0.75 mile in diameter. The tephra (ejected volcanic material) is a mixture of pumice ash and obsidian fragments and small pieces of lava, or lapilli. Within the crater is an obsidian dome. Pumice is widely distributed on the surface in this area as a result of the generations of eruptions from these craters. Indeed, just west of the craters is the aptly named Pumice Valley.

June Lake Loop

Where Highways 120 and 395 intersect, turn north (right), back toward Lee Vining. In about 0.5 mile, turn left onto the *June Lake Loop*. Four beautiful lakes are strung along this horseshoe-shaped valley: Grant, Silver, Gull, and June. The former is visible after about 3 miles from Highway 395.

Rush Creek leads up the valley to *Silver Lake*. Aspen and willow line the way, contrasting with the stark rock faces of the neighboring mountains. The lakes and the creeks offer excellent fishing. Support facilities strung along the road are most concentrated near the southern end of the loop, at *June Lake*.

Just south of Silver Lake, as the road begins to wind back toward the east, Carson Peak (10,909 feet) stands above the valley. Gull Lake appears next, and then you pass through the village of June Lake. The June Lake recreational area is geared toward all seasons. There are ski slopes, restaurants, fishing spots and hiking trails. The lower valley, Gull and June lakes, and Reversed Peak, can be put into perspective from the vantage point on Oh! Ridge just beyond June Lake.

The entire loop is about 16.3 miles, then the road rejoins Highway 395 for the southward journey. Jeffrey pine is abundant here. Volcanism through the eons has shaped the landscape. About 2.5 miles south on 395, after the end of the June Lake loop, stands Wilson Butte on the right, a rhyolite dome about 1,300 years old. The road then crosses a pumice flat over Deadman Summit. This name recalls a gold-mining—related murder on Deadman Creek in 1861.

In about another 1.5 miles *Obsidian Dome* is visible on the right. The obsidian, volcanic glass, is formed through the rapid cooling of volcanic flows. It was frequently worked by native peoples to form tools and weapons.

Convict Lake to Tom's Place

About 5 miles south of Mammoth Lake is the turnoff (to the right) to *Convict Lake*. This gem sits in the bottom of a glacial valley that has been partially dammed by lateral and terminal moraines. The road to Convict Lake crosses the deposits; note the variation in size of the stones in the moraines.

The name Convict Lake is further testament to the violent history of this part of the state. Last century, six convicts, part of a larger group of escapees from the Carson City penitentiary in Nevada, holed up in this canyon. A gunfight with the posse resulted in the death of a local merchant, Robert Morrison, and the escape of the convicts. Several of the escapees were recaptured shortly, and two were lynched on the trail back to Carson City in retribution for Morrison's death. A more recent tragedy involved several drowning deaths in an attempt to effect a rescue through broken ice. This is commemorated in the monument at the eastern end of the lake.

South of the lake is *Mount Morrison* (12,268 feet), and colorful Sevahah Cliff backs the head of the lake. Laurel Mountain forms the northern flank of this valley. Again, especially in the cliffs

MAMMOTH LAKES

Farther south is the (right) turn to *Mammoth Scenic Loop,* a worthwhile detour. The village of Mammoth Lakes can accommodate most needs of the traveler, including the desire for scenic splendor. The area also offers some of the best skiing available in California. Landscape delights include the columnar basalts at the Devil's Postpile National Monument, rainbow falls, and the earthquake fault. Excellent views of the Sierran wall appear across the lakes and through the trees.

High Sierra landscape—exfoliating granite at a roadcut with a wind-weathered conifer. Photograph by Robert A. Rundstrom.

along the western and northern edge of Convict Lake, we see the natural palette of metamorphosed sedimentary rocks. Here, the exposures include ancient sandstones, hornfels, and metacherts. Some of the oldest rocks in the Sierras, perhaps 500 million years old, occur in the metasedimentary rocks of Mount Morrison. The elevation of these former seabed deposits is powerful testimony of the tremendous tectonic forces that reshape the Earth.

After returning to Highway 395, turn right, and continue south. *Crowley Lake* will appear east of the highway after several miles. Crowley Lake (an artificial reservoir) lies in the lower portion of the Long Valley Caldera. A massive volcanic eruption about three quarters of a million years ago formed the caldera. These eruptions were responsible for the ejection of the Bishop tuff, an ash-fall deposit comprising about 500 cubic kilometers of material laid down over a period of only days (perhaps less). After the eruption, the magma chamber collapsed to form the caldera. The Bishop tuff covers in excess of 1,000 square kilometers and averages more than 100 meters in depth. Elements of the ashfall have been identified as far away as Nebraska. At present, the U.S. Geological Survey is monitoring seismicity and local elevations around Long Valley, particularly near Mammoth Lakes, in anticipation of the volcanism indicated by magma chamber swelling. In 1982, the U.S Geological Survey went so far as to issue a "notice of potential volcanic activity" for the area, much to the dismay of the tourist-reliant businesspeople in the region.

Just beyond the south end of the Lake Crowley recreation area is *Tom's Place*. Food, gas, groceries, and other amenities for all-season recreation are available here. Tom Yerby acquired this partly developed site in 1919, and it has been a local landmark since. Tom's Place marks the unofficial southern entrance to Mono County, and serves as a gateway to many recreational opportunities.

South of Tom's Place, Highway 395 begins the descent into the Owens Valley via Sherwin Grade. This grade is named after James Sherwin who built a toll road up this slope to service the Mono mines. Descending the grade, note outcrops of Bishop tuff, in pink, white, and brown, to the east (left), and morainal deposits to the

west, across the valley. About 3 miles below the summit, turn out
for a perspective on the northern Owens Valley.

Bishop to Big Pine

In the distance ahead, agricultural development around the city of
Bishop stands out green against the sere landscape. To the left are
outcrops of the volcanic tablelands, a plateau above the northern
end of the Owens Valley. The White Mountains form the eastern
rim of the valley, and Wheeler Crest is to the west, behind Round
Valley. Extensive moraine systems are visible along the flanks of
Pine Creek where it flows out of the Sierras. To the southwest is
the triangular peak of Mount Tom (13,649 feet), and just to the east
lie the much smaller Tungsten Hills, near the south of which is the
center of Round Valley. The Tungsten Hills take their name from
the ore mined in the vicinity. One of the largest tungsten mines in
the United States continues operation in upper Pine Creek Canyon
(out of sight from this point).

Continue south on Highway 395 to Bishop. This area was settled in 1861 by the Samuel Bishop family, who wanted the lush
pastorage for their cattle. The mining discoveries in the vicinity,
and the arrival of the railroad, contributed to the growth of this
regional center. *Bishop* is now the only incorporated town in Inyo
County, and it still serves as a regional center for most economic
and recreational activities.

Bishop has many motels and restaurants, gas stations and grocery stores, and most other shopping needs can also be met here.
Noteworthy is Schat's Bakery, downtown on Highway 395, the
home of Schat's Original Sheepherder Bread. This is a good place
to obtain picnic supplies for the journey south, especially if your
trip includes the White Mountains and the Ancient Bristlecone
Pine Forest. The only reasonable alternative is shopping in Big
Pine, which affords a wide range of amenities. No services are
available on the Bristlecone loop itself.

Leave Bishop, continuing south on Highway 395 toward Big
Pine. En route, you will pass the radio telescopes owned and

Road to Big Pine out of Death Valley. Photograph by Al Glass.

operated by the California Institute of Technology, visible to the east. This site was chosen because the Sierra Nevada and White Mountains screen out artificial signals.

Big Pine is about 15 miles south of Bishop. Highway 168, the route to the Ancient Bristlecone Pine Forest, joins Highway 395 just north of town. Note again that there are no travel-support facilities on Highway 168 between this junction and the Bristlecone Forest. The round-trip is about 50 miles, and heavy snowfalls occasionally close the road during winter. The road conditions are sign-posted at the junction. Food, water, and gasoline are readily available in Big Pine. Take Highway 168, Westgard Pass Road, east across the Owens River floodplain and up the flanks of the Waucoba Embayment, via alluvial fan surfaces. You will pass the

BRISTLECONE PINES

The bristlecone pine tree will not win any prizes for stature or girth. It is a dwarf relative to the redwood, or even the Monterey pine. It is twisted, asymmetrical, and stunted. However, the bristlecone has its own beauty, derived from the millennia spent in blasting wind, ice, and snow, and searing heat.

The bristlecone pines (*Pinus longaeva*) are the oldest living things in the world. The oldest of the old stand rooted in the limestone outcrops of the White Mountains. Additional stands are found near the peaks of other mountains in California, Arizona, Nevada, New Mexico, and Utah.

The climate, elevation, and soil in the *Ancient Bristlecone Pine Forest* seem inhospitable in the extreme. Precipitation averages about 10 inches per year, and annual temperatures average around freezing. However, winter lows and summer highs depart considerably from the average. The bristlecone is found at elevations between about 9,500 and 12,000 feet, above the range of most tree species. Finally, the dolomite (limestone) soils are very alkaline, quite thin, and generally inhospitable to plant life.

The discovery of the age of these trees occurred by accident, as the result of coring projects conducted to study tree rings as climatic indicators. As Dr. Edmund Schulman counted the tree rings, we may envision his growing surprise and perhaps disbelief: more than 4,000 years represented by the annual growth marks in cores from a living specimen. That specimen still stands—Pine Alpha in the *Schulman Grove.*

The Schulman Grove is the first of two dense stands of bristlecone pines found in this area. The oldest known trees are here, including the 4,600-year-old Methuselah Tree. There is also a small visitor's center and picnic area. Note, however, that no travel supplies are available. Several marked paths lead to some of the more remarkable trees. Be sure to differentiate between the bristlecone and the limber pine: the latter sports a green cone and its needles occur only in small tufts very near the ends of branches.

Bristlecone pine. Photograph by Martin S. Kenzer.

Many of the bristlecone pines look as if they are barely alive. Indeed, that is usually the case. These trees require that only a small proportion of their wood remain viable. Life is carried through the bark. As long as strips of the living tissue remain, the bristlecone can survive and reproduce. On most trees, the striking expanses of near-white, polished wood are eons old, dead wood.

The other main bristlecone stand is the *Patriarch Grove*. This area is about 12 miles from the Schulman Grove, over an unimproved road. The Patriarch Grove is at about 11,000 feet, and any physical exertion can be quite strenuous (this is also the case anywhere within the Ancient Bristlecone Pine Forest). The stand is named for the Patriarch Tree, the largest of the bristlecones. Millennia of wind and sun have blasted the trees here into an otherworldly configuration, making this place a photographer's delight. Please note that this is a very fragile ecosystem: Leave the trees (and other vegetation) for the next millennia.

turn toward the northern end of Death Valley, a long, largely unpaved route. The light-colored, dissected sedimentary rocks are ancient lakebed deposits.

After about 5 miles the road bed is constructed along the bed of a dry wash. In the lower wash, the alluvial deposits look very much like moraines, but they are not. In the upper reaches of the wash, the sheen of desert varnish covers exposed rock surfaces. Just over 8 miles from the Highway 395 junction is a historical interest site, Toll House Spring. The passage through these mountains was first built as a toll road to serve miners and prospectors in the White Mountains and Deep Springs districts. The elevation here is about 6,000 feet.

As the road climbs, the white rocks of limestone and dolomite outcrops become frequent. Some of these sedimentary deposits are as old as 600 million years. the dolomite outcrops provide a favorable soil for the bristlecone pines.

The road crests at *Cedar Flat Plateau*. Note that the trees here are in fact junipers. After about 1 mile, turn left on White Mountain Road, sign-posted for the Ancient Bristlecone Pine Forest. Over the next 10 miles, the road rises through several altitudinally zoned habitats, including those dominated by piñon pine and juniper. However, these trees give way to an alpine scrub vegetation, until the bristlecone and limber pine communities appear near the Schulman Grove.

Retrace your route back to Highway 395. The descent from the White Mountains offers some superb views of the Sierra Nevada and the Owens Valley. Be sure to stop at the Sierra View vista point, at about 9,000-feet elevation, and hike out to the overlook. To the east are the Grapevine Mountains on the far side of Death Valley. The Palisades Glaciers stand out clearly at the crest of the Sierras, and on a clear day Mount Whitney can be seen in the south, and Mount Dana, in Yosemite, in the north.

At the junction of Highway 168 with Highway 395, turn south (right) to *Big Pine*. This village is named for the now missing stand of pines that marked this spot. Big Pine stands below the *Palisades Glacier,* the southernmost glacier in the United States. Most travelers'

Cinder cones. Photo courtesy of National Geographic Society.

facilities are available in Big Pine, and there are excellent camping and fishing sites in the vicinity.

South of Big Pine, the Inyo Mountains now form the eastern rim of the Owens Valley. This is also the beginning of the Taboose–Big Pine volcanic field, active over the last 20,000 to 100,000 years. Crater Mountain appears just to the west of the highway. The light-colored granitic knob on its northern flank points to the geologic nature of Crater Mountain; a granitic plug overlain by darker basalt flows. Many cinder cones occur in this area, including the aptly named Red Mountain, visible to the south as you pass Crater Mountain. The line of eruptive activity roughly follows the faulting responsible for the 1872 earthquake that devastated much of the development in the Owens Valley.

Alabama Hills

About 11 miles south of Big Pine, large basalt flows can be seen to the west. On the steep Sierran slopes, the light gray of active screen slopes stands out. About 16 miles south of Big Pine is a roadside rest area where the highway rises to cross the base of the basalt flow. This is a good opportunity to observe details of a lava flow, especially the texture and composition of the rock. From this rise, the green of agriculture surrounding the settlement of Independence can be seen ahead. The *Alabama Hills* stand out near the center of the valley floor in the distance. The road rises and falls over large alluvial fans built out from the Sierras by mountain torrents.

Independence is about 27 miles south of Big Pine. The town name commemorates the establishment of a military settlement near this site on 4 July 1862. Fort Independence was to be the base from which the Indians of the Owens Valley could be subdued. Independence remains the Inyo County seat, despite its relatively small population. Sites worth seeing include the eastern California Museum, the county courthouse, and the Austin House. Back on Highway 395, head south toward Lone Pine.

It is difficult for the modern traveler to imagine that much of this part of the Owens Valley was once a richly productive agricultural region, before the city of Los Angeles purchased water rights and much of the land. However, about 4 miles south of Independence, views of lush alfalfa fields hint at what this landscape could look like with irrigation.

Approximately 2 miles south of *Manzanar*, Highway 395 curves to the left. At this location, you cross the Los Angeles aqueduct. When the road straightens again, you have a good view of the ancient-appearing Alabama Hills. The trees growing near the base of the hills roughly mark the 1872 earthquake scarp. Just as you pass these trees, the *Alabama Gates* can be seen over your right shoulder (west of the road). This section of the Los Angeles aqueduct was subjected to numerous dynamiting episodes during a local rebellion against the water transfers to the south. In November 1924, the Alabama Gates themselves were seized by a party of

MANZANAR

The stereotypical image of California as an open society whose citizens are fun-loving, carefree, and motivated to be tolerant of others can be misleading. In late March 1942, 10,200 Californians of Japanese descent, two thirds of whom were American citizens by birth, were removed at gunpoint from their homes and businesses and relocated to an austere camp in the Owens Valley called Manzanar. It was a wartime act born of racist hysteria on the West Coast, an action completely incompatible with the image most Californians have of themselves. For that reason the name Manzanar is one most Californians would like to forget.

Manzanar is in high desert country on U.S. Highway 395, squeezed between the highest peaks of the Sierra Nevada and the 11,000-foot Inyo Mountains. Drifting snow and subfreezing temperatures in winter and the searing 110 degree Fahrenheit heat in summer make the climate more like that of continental Nevada than California. Sagebrush and salt-tolerant shrubs are common. The name of a nearby town, Lone Pine, indicates the extent of local forest. The exotic Owens River, the main source of life here, had been appropriated by the city of Los Angeles in 1913. The name Manzanar derives from the Spanish *manzana,* or apple, an apt description for the abundant fruit production that occurred here before the water was taken. Today, Manzanar's most conspicuous features are two old stone guardhouses squatting in the wind under pagoda-like roofs.

A streak of anti-Asian sentiment runs continuously through California history, in stark contrast to the affinity for Asian culture also found here. With the Japanese defeat of the Russian navy in 1905, Californians began to see Japan as a dangerous world power, and a virulent racist campaign began against the local Japanese population. In short, many sought the same prohibitions with which the Chinese had been excluded since the 1880s. The attack on Pearl Harbor in December 1941 clearly inflamed racial tensions in California, and the "Civilian Exclusion Order" for "Japanese Aliens and Non-Aliens" came just four months later. Three conditions make this federal military order all the more noteworthy for California: Similar orders were not issued for Americans of German or Italian descent, nor for the hundred thousand Hawaiian Japanese, and most of the evacuees were American citizens whose individual constitutional rights were suspended.

Japanese-Americans born in the United States are called Nisei; those born in Japan are Issei. Only one third of the evacuees were Issei, all of whom had lived in the United States at least eighteen years. Ninety-two percent of Manzanar internees were farmers, merchants, shopkeepers, teachers and university students living in close-knit families in Los Angeles County when they received orders to evacuate the coast immediately, keeping only the belongings they could carry. Their land and homes were confiscated and sold off at reduced prices. Ultimately, not one of these evacuees would ever be convicted of spying or acts of insubordination.

Manzanar was built as a self-sustaining city, albeit one surrounded by barbed wire, with soldiers in guard towers armed with machine guns. Families were encouraged to stay together, and internees were allowed to print a newspaper, and practice their religion, trades, and professions. Nisei farmers were remarkably successful in this unforgiving landscape, and were in demand as temporary workers outside Manzanar. Of course, they were returned to the camp following harvest. Ironically, the surest way out of Manzanar was to serve in the U.S. Army, and many Manzanar men and women fought with distinction in Europe and the Pacific. In all, 33,000 Nisei fought in the U.S. Army in World War II, suffering a record 9,486 casualties.

Manzanar was closed in September 1945. With little to return to except "No Japs Wanted" signs, a generation of

formerly self-sufficient people returned to Southern California to lead broken and dependent lives. In 1988, each of the
few remaining survivors of Manzanar was finally awarded
$20,000 compensation by the federal government. Forty per-
cent now live in the Los Angeles area—one fifth of them
below the poverty level. Today, the names and dates on the
graffiti-covered walls inside the abandoned guardhouses
mark Manzanar as pilgrimage site for the children and grand-
children of the Nisci.

armed protesters who opened them to return the flow to the Owens
River.

Shortly after the Alabama Gates, the road drops down over the
1872 earthquake scarp. The 1872 earthquake was the largest in
California's recorded history, 8.3 on the Richter scale. Most buildings in the area were destroyed or damaged heavily, twenty-seven
persons were killed—a substantial proportion of the population at
that time. The road follows the base of the earthquake scarp for
several miles. The mass grave site of sixteen of the victims is
marked by a historical plaque, also to the west of the highway, just
north of Lone Pine.

Lone Pine

Lone Pine was settled in the early 1800s, and the town developed
largely as a regional supply center. It remains a focal point for
southern Inyo County, and a gateway to Mount Whitney and Death
Valley. The area offers excellent fishing, backpacking, and camping along the eastern Sierra slopes. The town has most amenities of
interest to travelers, including restaurants, grocery and sporting
goods stores, lodging, and gas stations.

At the one stop light in Lone Pine, turn right (west) onto Whitney Portal Road. This road leads high on the flank of the moun-

tains and presents grand views over the Owens Valley and Owens Lake bed. This route takes you through the Alabama Hills, named after a Confederate warship. These hills were once thought to be of great age, but in fact they are much younger than the Sierra Nevada behind them. The odd shapes of the rocks results from spheroidal weathering that occurred while the rocks were covered by overburden.

About 2 miles out of Lone Pine, turn right onto Movie Flat Road. As you drive along this route, look for the distinctive scenery that has made this region a favorite of film-makers and television producers. These foothills have been used for productions including *Gunga Din, Lives of a Bengal Lancer, Along the Great Divide,* "Cimarron Strip," "Rawhide," "Bonanza," "Wagon Train," and "The Lone Ranger."

After your cruise through these strange hills, return to Lone Pine and park near the traffic light. On the southwest corner of the intersection is a pair of footprints painted onto the sidewalk. If you stand in these prints and look Sierra-ward, *Mount Whitney* will be directly before you. At 14,495 feet, this is the highest mountain in the contiguous United States.

Leave Lone Pine, travel south about 1.5 miles on Highway 395, and turn east onto Highway 136. An immediate right turn brings you into the parking lot of the U.S. Interagency Visitor Center. This is a good conceptual beginning for the Death Valley segment of your journey. The visitor's center sells a wide variety of pamphlets and books describing not only Death Valley, but the eastern Sierra–Owens Valley in general. You can also obtain the latest information concerning roads and weather at this location.

Leaving the parking lot, turn east (right) onto Highway 136. This route takes you first across the Owens Valley. You will cross the Owens River, in its presently diminished state, near the middle of the valley. To the right (south) can be seen the expanse of Owens Lake bed. In the left distance, new-looking flow marks can be seen on the alluvial fans spreading from the Inyo Mountains. These recent scars are at least a decade old, evidence of the extremely slow rate of change in this arid environment. The white mine tailings near these fans are remnants of dolomite mining.

Curving ridges mark the old shorelines of Owens Lake, north end.
Photograph by Bernard O. Bauer.

Keeler

About 10 miles from the visitor's center, note the small sand-dune fields on the right. These dunes were derived from the beach sands associated with the higher levels of Owens Lake that occurred before the diversion of Owens River water to the Los Angeles aqueduct. After another 3.5 miles, turn right on Malone Street, and enter Keeler.

Keeler once served as the supply-line gateway to the rich silver–lead mines of Cerro Gordo. Indeed, Keeler was a port city, the landing site for steamers plying Owens Lake. The steamers brought charcoal, foodstuffs, and other supplies to the mines and smelters, returning to Cartago on the opposite shore with the more precious, metallic cargo destined for Los Angeles.

Creosote bush, Panamint Valley. Photograph by Paul F. Starrs.

Present-day Keeler is still a community based around a mining and prospecting population, but it is also becoming a center for retirees. Follow Malone into Keeler. You will cross Railroad Avenue, with the old station on the right. At the end of the street turn left. This road roughly parallels the edge of town when the lake levels were deep enough to support the steam vessels.

Turn left on Cerro Gordo Street. On the left is the old town school, and to the right is the Sierra Talc Company Mill. Note the relatively high-density housing within the townsite, compared with the open spaces of the surrounding desert. Turn right, back onto Highway 136 at the edge of town.

About 0.5 mile east of Keeler is the turn (left) to *Cerro Gordo*. Storms make this road impassable at times, as do changing property responsibilities; the road was closed in summer 1991. Along this stretch of Highway 136, ruins of structures associated with mining, smelting, and reducing mineral products abound. Many of them are still used, despite the often dilapidated appearances.

Joshua tree. Photograph by Elliot McIntire.

Five miles from Keeler is the junction of Highways 136 and
190. Take Highway 190 east, sign-posted for Death Valley. Note
the shadow dunes, mainly along the right side of the highway.
These dunes are sands deposited in the lee "shadows" of other
landforms. The next several miles evince the area's volcanic past,
including a large basalt flow from the left and ash deposits. About
6 miles from the junction, widely spaced Joshua trees appear on
some of the alluvial fans. The Joshua tree is environmentally
sensitive, and this is one of very few stands outside the Joshua Tree
National Monument.

Darwin and Panamint Valley

The road to Darwin turns off to the south about 13 miles from the
Highway 136/190 junction. Follow this road for about 5 miles into
the small mining town. *Darwin* is named after Dr. Darwin French,
who, along with several companies of prospectors, made several
trips through this region searching for gold. They found silver and
lead in this district, where some mines still operate.

Several generations of company housing tracts can be seen in
this village. On the hills to your left is the most modern variety,
with most units obviously suited to accommodate bachelors. In the
town proper, similar ranks of uniform, but even smaller housing
still stand. Remember that in environments like this, almost all
construction materials other than rock and sand had to be freighted
great distances across rugged desert terrain—in this case, usually
from Keeler.

At the T junction in town, turn right onto Main, down the hill.
Turn right on Fulton to see some vernacular landscaping, then
U-turn back to Main and continue down the road. Please note that
most of Darwin is still private property, and many of the houses
that look abandoned in fact are not. At the base of Main, a small
dirt road continues down the hill into the gulches below town.
Here also stand several reminders of how difficult life in Darwin
was. Rows of holes can be seen, dug into the sides of several of the
gully banks along the wash. These are remnants of houses dug by

those who could not afford the luxury of wood or sheet metal. One advantage to these partial cave-dwellings is that the insulating properties of the earth modulated the extremes of temperature common to the high desert. Note also that there are some facilities for the traveler in Darwin, including a museum that seldom seems open. After your tour, retrace your tracks back to Highway 190, and continue east toward Death Valley.

About 4 miles from the Darwin turnoff, you will pass a road sign-posted for Saline Valley and Big Pine. This is one of the end points for the turnoff passed on Highway 168 en route to the White Mountains. After another mile, as you crest a rise, the Panamint Mountains will be visible to the east.

Ten miles from the Darwin junction, turn off at the Father Crowley Vista Point. The Very Reverend Monsignor John J. Crowley was also known as the Desert Padre; Crowley Lake also bears his name. Proceed past the first, large parking lot on the dirt road that trails toward the east. To the left is a deeply eroded gorge, cut through a basalt flow into much softer sediments beneath. At the turn-around at the end of the dirt road, stop for the view of the *Panamint Valley*, named after the native tribe that once inhabited what appears now as a wasteland. The Panamint Mountains, to the east, display evidence of the large-scale faulting and folding responsible for the contorted landscape in this region. Beds of sedimentary rocks can be seen twisted at right angles by tectonic forces. Note also the large sand-dune field at the northern end of the valley, trapped by the rising topography. A red cinder cone can be seen on the slopes below, and the scattering of black and red cinders across the hills also attests to volcanic activity through this region. The southern and central sections of Panamint Valley are dominated by a dry lake bed. After you have seen enough, return to Highway 190 east, down the winding grade.

About 8 miles from the *Father Crowley Vista,* you pass the Panamint Springs resort. Gas, food, and lodging can be had here. Note again the star-shaped dunes to the north. Continue across the valley floor and up the Panamint slopes. The various colored rocks near the summit attest to the diverse geological history of this mountain range, where abutting morphosedimentary assemblages and volcanic deposits have both been contorted by intensive faulting and folding.

Death Valley

Highway 190 crests the Panamints through *Townes Pass* (4,956 feet), the western entrance to Death Valley. From Townes Pass, the road descends steeply over an alluvial fan surface toward the heart of the basin. About 6.5 miles after the summit, the 3,000-foot elevation is crossed, and Death Valley proper opens to view. The Grapevine Mountains, visible earlier from the White Mountain excursion, form the distant rim of the valley.

Stove Pipe Wells is about 17 miles from Townes Pass. Facilities include gas, food, and lodging. There is also a swimming pool, open to the public for a modest fee. The original Stove Pipe Wells Hotel was built in 1926 by the Eichbaums to serve travelers entering the region over Townes Pass via Eichbaum's toll road. The latter has evolved into present-day Highway 190, and the resort is now owned by the National Park Service. The large trees near the village are tamarisk, introduced from the Mediterranean. Stove Pipe Wells is the unofficial hub for exploring the western and northern reaches of Death Valley National Monument.

Just east of Stove Pipe Wells, Highway 190 drops below sea level. *Death Valley National Monument,* established in 1933, includes the lowest elevation, –282 feet, in North America. Structurally, Death Valley is a graben—a block of the Earth's crust that has dropped between adjacent structures. The block is also tilted from the west to east. The most obvious manifestation of the tilting in the southern basin is the discrepancy between the size of alluvial fans on opposing sides of the valley. Massive fans spreading from the Panamint Range are obvious along almost the entire western flank of the graben. The eastern edge has a few large fans, as the sediments debouch onto a slowly sinking surface.

After another 5 miles, the *Devil's Cornfield* bounds Highway 190. Tall arrowweed grasses trap and anchor sediments against the eroding wind, locally raising small hillocks. The arrowweed ob-

"Land yachts" carry tourists in comfort through the Death Valley heat, Stovepipe Wells. Photograph by Paul F. Starrs.

tains water from seepage out of Triangle Spring. Immediately north of the Devil's Cornfield is a large dunefield. This extensive sand body presents numerous photo opportunities, especially in the low-angle light hours of morning and evening.

About 3 miles past the Devil's Cornfield, Highway 190 intersects with the roads leading north to Scotty's Castle, Ubehebe Crater, and Rhyolite and Beatty via Mud Canyon. The highway describes a series of broad curves around the bases of alluvial fans. After about 10 miles, broad salt flats are visible to the right, continuing into the far distance in the bottom of the valley.

Another 6 miles brings you to the exposures of yellow sedimentary rocks in the vicinity of Mustard Canyon, mainly to the east. The old *Harmony Borax* works are sign-posted to the right of the road. This facility was built by William Coleman, who purchased the rights to the borax claim from Aaron Winters in the 1880s. The borate deposits were concentrated at the surface by evaporative processes. It was the delivery of borax that gave rise to the famous twenty-mule teams, used to freight the salts across 165 miles of desert to Mojave. As you stand near the ruins of this enterprise, imagine working in the searing heat to break chunks of the mineral free from the surface of the salt pan, or tending the steaming boilers to reduce the borates to a purer state. Then get back into the air conditioning for the journey to Furnace Creek.

Two miles beyond the old borax works is the modern *Furnace Creek* Visitor's Center. This is a good stop for information and literature concerning this region. Helpful and friendly National Park Service staff will assist to make your tour more enjoyable, and safer. One-half mile farther brings you to Furnace Creek Ranch.

Furnace Creek Ranch is the largest development in Death Valley. Amenities include gas, food, lodging, a motel, and camping. This is also the site of the old Greenland Ranch, spreading across the alluvial fan from Furnace Creek Wash. Here was recorded the highest air temperature in the western hemisphere: 134 degrees Fahrenheit. Average high and low temperatures are about 65 degrees and and 39 degrees, respectively, in January, rising to 116 degrees and 88 degrees, respectively, in July. Rainfall averages less than 2 inches per year. These statistics make Furnace Creek both the hottest and driest location in the United States.

Death Valley to San Fernando Valley

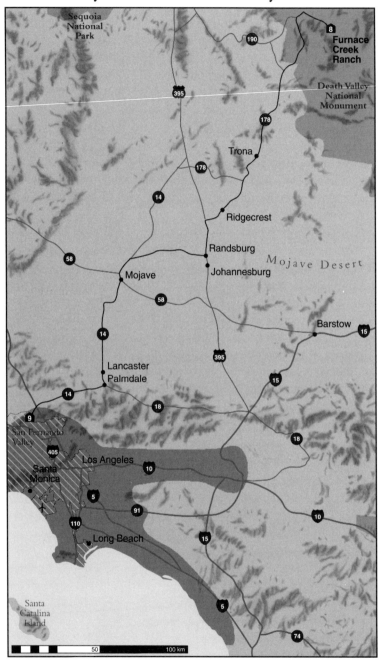

△ *Day Nine*

DEATH VALLEY TO SAN FERNANDO VALLEY

The final day's itinerary takes you from some of the most isolated territory in North America to cities overflowing with people and cars, from the desert landscapes of Death Valley to the artificially watered suburbs and neighborhoods of Los Angeles. You begin by exploring still deeper into Death Valley, then retracing your trip back past Stove Pipe Wells and then heading southwest toward Palmdale en route to the "City of Angels."

Zabriskie Point and Dante's View

Leave Furnace Creek Ranch, turning right (south) on Highway 190. In less than 1 mile, at just about sea level, the road to *Badwater* (Highway 178) forks to the right. Continue on Highway 190. Furnace Creek Inn is on your left. This imposing structure was built originally to serve the employees of the Pacific Coast Borax Company. Now anyone with enough money can enjoy the splendor of this resort oasis—at least from October through May.

The outbuildings of the Furnace Creek Inn are built around the mouth of Furnace Creek Wash. Highway 190 follows the wash up out of the valley. Patches of vegetation grow along the road, on the floor of the wash, and in hollows to the east. The greenery marks a series of springs that feed into the alluvial fan under Furnace Creek Ranch.

About 3 miles beyond Furnace Creek Inn, turn right (sign-posted) into the parking lot for *Zabriskie Point*. A short, steep hike leads to an overlook of Death Valley. Here, the sharply contrasting geologic formations provide endless opportunities for the photographer and the observer. Colors change slowly through the day as the sun highlights first one formation and then another. Early morning or late afternoon affords the best viewing, as shadows accent the warm tones of the landscape. You will also see Manley Beacon, an arced finger of golden sedimentary rock pointing across central Death Valley.

Northeast of the Zabriskie Point overlook, note the incised channel in Furnace Creek Wash. It looks like a river channel that stops at the upper edge of the parking lot. Just below you is an artificial cut in the soft sedimentary rocks that diverts flow out of the wash and into Gower Gulch. Diverted sands and gravels line the bed of upper Gower Gulch. The more adventuresome may want to enter the diversion, from the parking lot, and examine the scoured channel cut by fifty years of flood waters.

Leaving Zabriskie Point, continue up the wash on Highway 190. Note the sharp contrast in surface materials between the east and west sides of the road. To the west, ancient lake-bottom sediments comprise the brightly hued badland topography. To the east lie scattered igneous rocks from ancient vulcanism.

About 6 miles from Zabriskie Point, is the right turn from Highway 190, at *Ryan Junction*. Take the turn (sign-posted) to Dante's View. On the hillside directly before you are the tin-roofed remains of Ryan. This used to be a railroad town and support facility for borax mines in the region. It is still private property. Just before you pass Ryan, the Billie Mine is on the left, and extensive mine tailings are on the right. Both are also remnants of borax extraction.

The road to *Dante's View* continues uphill, past Greenwater Canyon, and into the Black Mountains. About 12 miles after Ryan Junction, the road climbs steeply to the parking lot and overview. Note that places are provided to leave trailers or large recreational vehicles near the summit; the last quarter mile is tightly hairpinned and very steep.

Zabriskie Point, Death Valley. Photograph by Elliot McIntire.

Badwater. The lowest point in Death Valley—282 feet below sea level. Photograph by Elliot McIntire.

Dante's View is more than 1 mile above sea level (5,475 feet). In Death Valley, directly below, lies Badwater, the lowest point in the United States, at –282 feet. Across the valley, Telescope Peak crowns the Panamint Range, at 11,049 feet. Because of the elevation at Dante's View, temperatures are usually pleasant in summer, and can be quite cold in winter. Well-marked displays explain cultural and physical aspects of the view before you. Note, especially, the appearance of desolation as you scan south and north along the axis of the valley. Imagine the despair of the '49ers as they braved the long crossing in ox-drawn wagons. On the west side of the valley, thin lines of green vegetation indicate springs brought to the surface by faulting. Many of these springs were attractions for the native Indians, early explorers, prospectors, and settlers. Try also to imagine the extent of Lake Manley, about 10,000 years ago, as it filled around the south end of the Panamint Mountains.

BORAX

It is one of the many ironies of Death Valley exploration and prospecting that great mineral wealth, often sought for at great cost in life and fortune, came not in the form of precious metals like gold or silver, but from the common evaporate salts glistening white on the valley bottom. Sodium tetraborate, borax, spawned one of the most dramatic eras in the development of Death Valley.

Throughout most of the eighteenth and nineteenth centuries, the world's supply of borax was furnished, thinly, from remote deposits in Tibet and from scattered sources in Chile, Turkey, and the American West. The commercial applications of this mineral grew faster than the supply. Borax was used as a meat preservative, in the production of lenses, in ceramics and pottery, as a rust-inhibitor, weed killer, mouthwash, and as a laundry product. However, it remained an expensive additive because of the poverty of sources.

In 1881, Aaron and Rosie Winters were hard-scratch farmers struggling to subsist on the Nevada side of the Funeral Mountains when fortune brought Harry Spiller to their door. Spiller shared with them stories of prospecting and hard times, and, in particular, details of his search for borax. A some-time prospector himself, Aaron Winters recognized the "cottonball" samples carried by Spiller. Winters had seen lots of the stuff in nearby Death Valley.

The next day, near Furnace Creek, the Winters used the borax test described to them by Spiller. According to the story, Aaron shouted "She burns green, Rosie! By God we're rich!" The Winters sold the claim to the W. T. Coleman Company for $20,000, and the rush was on.

Coleman and his partner, F. M. Smith, developed the borax works in the gut of Death Valley. Chinese laborers hacked roads, wagon trains struggled to bring parts and supplies, and the cottonball was gathered from the white Hell of the salt flats. The ruins of the Harmony Borax works are reminders of what men went through for cleaner laundry. Standing near the old boilers, one cannot help but empathize with the human effort required to get water, fuel, food, any machine part, even labor.

In addition to the local obstacles, there remained the problem of taking large bulks of borax to markets. The nearest railroad was 165 miles away, at Mojave. There was no road, no settlements, and water was scarce and widely spaced. The answer was the construction of giant wagons, weighing almost four tons, with wheels seven feet high and beds 16 feet long. Pairs of these wagons, joined with a water-wagon, were pulled across the desert and out of Death Valley by teams of twenty mules. The twenty-mule teams went on to become famous symbols of borax and of human endeavor. The depiction of the mule trains graced many a laundry shelf in the twentieth century. Some of the original wagons remain, and they can be seen at Harmony Borax Works and Furnace Creek Ranch.

The romance of grubbing cottonball from the bed of Death Valley died with the discovery of "colmanite" deposits in the local mountains. Colmanite was easier to process than cottonball, and the deposits were marginally closer to railheads. Large deposits of borate of soda were later discovered at (present-day) Boron, California, greatly increasing borax production, and again lowering prices.

Morning sun especially highlights the distant relief and points up the subtleties of the desert landscape. Return to Highway 190 and follow it back down the wash toward Furnace Creek Inn.

After you pass the inn, turn left onto Highway 178, toward Badwater. Almost immediately, a small earthquake scarp is visible to the left of the road. It is identifiable as a linear trace across

PLEISTOCENE GHOST LAKES

About 1.8 million years to about 10,000 years ago, during the Pleistocene Epoch, the Earth had its most recent Ice Age. The average temperature of the earth was lower than present, and many areas experienced increased precipitation and growth of glaciers, ice caps, and ice sheets. The high mountain ranges of California, including the Sierra Nevada, Inyo, White, and Panamints, were ideally configured to catch and store (in the form of snow and ice) the increased precipitation because of their north–south orientation. Probably only the Sierra Nevada had semi-permanent glacier systems. Each spring would send torrents of melt water flowing into the basin valleys. Furthermore, cooler temperature lowered evaporation rates from the open waters, enhancing the residence time of water in the lakes. Finally, as the climate warmed, large-scale melting of glacial ice led to additional flooding from the mountains into the Pleistocene lakes system. This large-scale flooding happened more than once during the Pleistocene, as temperatures rose and fell in slow cycles lasting more than 100,000 years.

Lake Russell (present-day Mono Lake) was once, geologists believe, part of the drainage system entering the Owens River until the eruption of the Long Valley Caldera about three quarters of a million years ago (about mid-way through the Pleistocene).

Owens River was a substantial stream during the Ice Age, delivering large volumes of fresh water to Lake Owens (now Owens Lake). The lake was about 250 feet deep, filling the broad basin and overflowing toward the south. These waters moved to create a shallow body of water, ancient Lake China (presently China Lake). This lake was perhaps 40 feet deep, and overflowed to the east into Lake Searles, a larger and much deeper pool. During periods of very high water, Lakes

China and Searles formed a continuous body of water. The combined surface area of these lakes was more than 350 square miles.

At its maximum, Lake Searles was more than 600 feet deep, and filled its basin entirely. Relics of the high water include ancient shoreline features and the tufa pinnacles at the southern margin of the basin. Lake Searles overflowed around the southern end of the Slate Range, into the Panamint Valley to fill Lake Panamint. Lake Panamint would have been an impressive site. The maximum lake area would have covered about 60 miles of the valley length, with widths up to about 10 miles. Maximum water depths would have exceeded 900 feet. At its high stands, Lake Panamint would have crested modern Wingate Pass to flow into Death Valley. The floor of Death Valley was occupied by Lake Manley, 90 miles long, 10 miles wide, and 600 feet deep. Lake Manley was the end of the line for Pleistocene run-off, however. There is no evidence for the filling of the Death Valley basin to overflowing.

The evidence for high water levels is abundant for all of these Pleistocene lakes, although it is hard to imagine their presence today. It has been more than 2,000 years since substantial waters have occupied any of the southern basins. The remnant salt and mineral deposits lining the dry lake beds are arguably the most notable bequest of these lakes.

alluvium on some of the small fan surfaces. Continue along this road to *Badwater,* about 19 miles from the turnoff. There is a parking lot and a path to lead you out onto the salt flats where you could be blistered by the sun and, without proper protection from the sun, die a hideous death. While in Badwater, look at the mountain behind you and note the sea-level marker—nearly 300 feet *above* you.

From Badwater, return north to Furnace Creek Ranch and continue back along Highway 190 north and then westward past Stove

Pipe Wells. Take Highway 178 south toward Ballarat, Trona, and Mojave. As you go through this astonishingly dry land, imagine what it was like for the borax miners and the mule-trains that pulled the wagon-loads of borax all the way to Mojave.

About thirty miles south of the intersection of Highway 190 and Highway 178, take the side road on your left eastward to Ballarat. This side road will rejoin Highway 178 about 10 miles farther south after passing the salt bed of a former lake. Continue on Highway 178 south another 30 miles to Trona, passing by Searles Lake to the east. These lake beds (like Mono Lake and Owens Lake) are evidence that the Basin Ranges province was once rich in the melted water from ancient glaciers.

These barren, unpopulated lands are now prized most for their isolation: the perfect place to test weapons without argument from nearby citizens. Most of the land on either side of Highway 178 is used by the military for ordinance testing, and unauthorized visitors are not allowed.

About 1.5 miles south of Trona, watch for a poorly marked dirt road on your left to the tufa pinnacles left behind when Lake Searles dried up. The extent of the area with pinnacles makes you realize how large the lake once was.

Follow Highway 178 from Trona to Ridgecrest; then join Highway 395 south (left) to the gold and silver mining towns of Johannesburg and Randsburg. From Randsburg follow signs for Highway 14—it is now about fifty miles on to Mojave. You will see the southern end of the Sierra Nevada off to the west. Continue south to Palmdale.

Palmdale

Palmdale has its origin in the founding of two separate towns during the late nineteenth century, Harold and Palmenthal. Harold was a small railroad town established by the Southern Pacific in 1885. Palmenthal was inaugurated in 1886 by Swiss and German families, predominantly from Nebraska and Illinois. As the story goes, they were advised that when they saw palm trees they would

Central Valley Project aqueduct along Highway 5. Photograph by Paul F. Starrs.

be close to the Pacific Ocean. As they came through the Antelope Valley, they evidently mistook Joshua trees for palm trees, and they settled the area south of the present-day city of Palmdale.

The early 1890s was a boom period for the newcomers; cheap land and tremendous agricultural potential were the main inducements. During the later 1890s, drought brought about agricultural failure, and many of the original settlers elected to abandon their farms for more lucrative opportunities elsewhere. Both Harold and Palmenthal were renamed in 1899 and relocated to the present site of Palmdale.

Water was to remain a serious problem for the inhabitants of Palmdale for a number of years. But, with the completion of the Los Angeles aqueduct, in 1914, and with the availability of electricity, agriculture finally took firm hold and became the primary means of livelihood in the area. With access to water, Palmdale's population began to multiply steadily. Irrigated lands in the valley mushroomed from 5,000 acres in 1910 to 11,900 in 1919. Alfalfa, pears, and apples soon became staple crops across the valley.

Agriculture remained the primary industry of the Palmdale area until World War II. With the coming of the war, however, Palmdale promptly turned into an important aircraft assembly and testing area. Following the war, Edwards Air Force Base and Air Force Plant 42 became notable as government defense, research, and testing facilities. The city continued to grow (as part of L.A. County), until it was decided in 1962 to incorporate. By the late 1970s, Palmdale was referred to as the aerospace capital of the United States: Rockwell, Northrop, Lockheed, and McDonnell Douglas all maintained facilities in the immediate area. The B-1 and B-2 bombers were manufactured in Air Force Plant 42. In addition, the Federal Aviation Administration's Air Route Traffic Control Center, which handles all air traffic for its Western Region, is also stationed in Palmdale.

Growth throughout the region has been extraordinary over the last couple of decades. In fact, Palmdale proved to be the fastest growing city in California between 1980 and 1990—spiraling during that 10-year period from 12,227 to 68,842—an increase of 573 percent! Recent population growth in the area is unrelated to industrial or agricultural growth, however; mostly it has to do with affordable housing in the L.A. metropolitan area. Today, Palmdale has effectively extended 10 miles north, to the point where it has practically come in contact with the city of Lancaster, and distinguishing between the two suburbs has become rather difficult. With house prices soaring in L.A., Palmdale and Lancaster, in the course of little more than a decade, have become sprawling bedroom communities, with an increasing number of residents commuting into the central city to work—60 to 70 miles to the south.

The San Fernando Valley

Follow Highway 14 over Soledad Pass to Newhall and San Fernando. The *San Fernando Valley*—better known throughout Southern California as "the Valley"—constitutes one of the largest concentrations of population in the United States. Since the Valley is not entirely enclosed on all sides, its precise boundaries and population size vary. Based on conservative measures, however, the Valley can be said to run approximately 25 miles from Burbank to Chatsworth and to extend, at its widest point, 11 miles across. Within this small geographical area the 1990 census identified about 1.4 million residents.

Between 1980 and 1990 the population of the Valley increased by more than 11 percent, making it one of the more rapidly expanding urban areas in California. In fact, if all U.S. cities were ranked according to size, and if the San Fernando Valley were considered a separate city (rather than part of the greater Los Angeles area), it would rank as the sixth largest city in the country. Only New York, Chicago, Los Angeles, Philadelphia, and Houston would contain total populations larger than the Valley's.

The San Fernando Valley has, until recently, been viewed as an area of exclusively suburban sprawl. Traditionally the Valley has been thought of as a place comprised mostly of large housing developments aimed at middle-class families. More specifically, the homes were earmarked for young, middle-class families that derived their incomes from employment "over the hill," in Los Angeles proper. Rapid population growth during the 1980s, however, has transformed the Valley from an area of predominately white, middle-income, suburban sprawl to a medley of ethnic enclaves. During the 1980s much of the population growth resulted from a large influx of immigrants from Asia, Mexico, South and Central America, and the Middle East. As a result, immigration combined with "white flight" out to the new suburbs on and beyond the edge of the San Fernando Valley has left the Valley with an almost 47 percent nonwhite population base. Much of this ethnic change has taken place in the eastern portions of the Valley, in which some neighborhoods are almost exclusively ethnic in composition today.

Along with the new ethnic makeup of the Valley, the economic structure has changed, too. Currently, almost 70 percent of the Valley's residents work locally, rather than travel into Los Angeles for jobs. Electronic manufacturing, once a stable employment base of the Valley, has begun to shift westward, into Ventura County, along with numerous technologically oriented companies. Currently, employment in the Valley is 49 percent service-based, an overall increase of almost 13 percent during the past ten years.

While the Valley no longer fits the stereotypical suburban image it possessed during the 1960s, '70s, and '80s, it remains an important economic base for all of Southern California. With two significant aircraft companies, three movie studios, and two major high-rise office complexes (in Encino and Woodland Hills), along with the growing ethnic population, the San Fernando Valley has emerged as far more than a bedroom community servicing downtown Los Angeles. Today, the Valley is swiftly maturing into an important urban place in its own right.

Now join the commuters on Interstate 405 and continue south to Los Angeles and the Pacific coast.

CALIFORNIA—NO LONGER THE FUTURE?

As America's center of gravity continues to shift west, California is no longer the crystal ball of national trends it once was. Now, it is almost a mirror, or perhaps a simulcast, of the most important things starting to take place in the United States. Forget the tax revolts, the hot tubs, the sports cars, and the computers; the California of the 1990s will have to justify expensive real estate a few miles from impoverished ghettos; balance environmental resources with increasing population; and rationalize ownership of major corporations by Pacific Basin countries.

To some extent, the bloom is off the orange groves—California's culture is changing. The descendants of the Dust Bowlers who fled to California sixty years ago are moving back to Oklahoma and Texas. Today, so many Asians and Latinos are pouring into California that in two more generations the Golden State might take on a Third World coat. California is no longer a symbol of the future because, as the farm belt and rust belt go the way of the *Saturday Evening Post,* the state is fast becoming the "big today": California represents roughly 13 percent of the United States' gross national product and 12 percent of its population.

Multiracialism is another California enigma. With the number of Asians, Latinos, and blacks increasing every year, whites are fleeing to the suburbs and even out of the state. There is currently a net emigration of the white middle class, something California has never experienced before. Those departing the state leave behind a growing homeless population, a widening rich–poor cleavage, and an increasing tension between ethnic groups—witness the May 1992 riots in South-Central Los Angeles.

Still another California issue is an ever-rising sense of U.S. economic nationalism. The great debate no longer involves closed Appalachian steel mills. The new front lines of trade and investment pass through California: through the Silicon Valley, through the rice fields of the Sacramento Valley, and through Hollywood. The large cities where foreigners own half of the downtown buildings are not New York or Boston, but rather Los Angeles. California is also the place where Japan owns almost 30 percent of the local bank deposits.

The size of California thus makes it a high-intensity focus for the rest of the country. Its challenges in the 1990s will most likely become the nation's challenges. The surfboards and sports clothes are yesterday's trend-setting fads. The role California will play in the 1990s will be new, different, and its greatest yet!

PART THREE

Resources

△ Hints to the Traveler

California is a huge state and full of cars. No matter how long you think it will take to drive from point A to point B, allow at least twice that much time. Construction, rush-hour, pedestrians, logging trucks, snow, landslides, even delays from film crews might easily stop the flow of traffic somewhere along this route. Although we have tried to give clear directions, the roads may be blocked or made one-way unexpectedly. Call ahead to make sure that attractions are open.

In the more desolate parts of this itinerary, it is wise to fill up on gasoline at every opportunity and to carry plenty of water for your car and yourself. If you have a breakdown, you are safer staying with your car and in the shade it makes than wandering away for help.

California's mild weather is one of its attractions, but the sunshine can hurt. Put on plenty of sunscreen in all seasons. In Southern California and in the dry hot lands along the Nevada border, you will probably want to wear light clothing most of the time (except in air-conditioned buildings). Along the coast and in the Bay Area, though, the fog and wind can quickly force you into sweaters, coats, and scarves.

Poison oak (*Rhus radicans*) is a common weed in disturbed areas throughout much of California. Its shiny green and red leaves tempt the unwary into picking it. Don't! Get to know its distinctive pattern of three leaflets and avoid it. The oil (related to lacquer) in every part of the plant can raise itchy, painful blisters a few days after you touch the plant. It is either identical to or closely related to the eastern poison ivy (taxonomists disagree), but causes at least as much misery.

Lyme disease, a bacterial illness carried by ticks, which are in turn carried by mammals and—in California—also by lizards, has been recognized recently as a hazard of walking in grassy and brushy areas in many parts of North America. Your best protection is to wear ankle-length pants, tuck them into your socks, and check your skin for the tiny ticks afterward.

We have not included metric conversions of distance and length in the text. For non-U.S. travelers, however, 1 mile is equal to about 1.6 kilometers, or 1 kilometer equals 0.62 mile; 1 foot is equal to about 0.3 meter, or 1 meter is equal to 3.3 feet.

△ Suggested Readings

Armor, John and Wright, Peter (1988). *Manzanar.* New York: Vintage
Books (with photographs by Ansel Adams).

Bakker, Elna S. (1971). *An Island Called California: An Ecological
Introduction to Its Natural Communities.* Berkeley: University of California Press.

California Coastal Commission (1983). *California Coastal Access
Guide.* Berkeley: Univesity of California Press.

——— (1987). *California Coastal Resource Guide.* Berkeley: University of California Press.

Hart, James D. (1978). *A Companion to California.* New York: Oxford
University Press.

Hill, M. (1975). *Geology of the Sierra Nevada.* Berkeley: University of
California Press.

Hornbeck, David (1983). *California Patterns: A Geographical and Historical Atlas.* Palo Alto, Calif.: Mayfield Publishing Company.

Nash, Roderick (1977). *Wilderness and the American Mind.* New Haven:
Yale University Press.

Norris, R. M. and Webb, R. W. (1990). *Geology of California.* 2d ed.
New York: John Wiley and Sons.

Sharp, R. P. (1976). *Geology Field Guide to Southern California.* Rev.
ed. Dubuque, Iowa: Kendall/Hunt.

Starr, Kevin (1973). *Americans and the California Dream, 1850–1915.*
New York: Oxford University Press.

Starr, Kevin (1985). *Inventing the Dream: California Through the Pro-
gressive Era.* New York: Oxford University Press.

Starr, Kevin (1990). *Material Dreams: Southern California Through the
1920s.* New York: Oxford University Press.

Stewart, George R. (1971). *Ordeal by Hunger.* New York: Pocket Books.

Wilkinson, Nancy L. (1991). "No Holier Temple: Responses to Hodel's Hetch Hetchy Proposal. " *Landscape* 31 (Spring): 5.

△ Index

Note: All locations are within the state of California unless otherwise indicated.